Quantum Theory of Anharmonic Effects in Molecules

Quantum Theory of Anharmonic Effects in Molecules

Konstantin V. Kazakov

ELSEVIER

AMSTERDAM • BOSTON • HEIDELBERG • LONDON • NEW YORK • OXFORD
PARIS • SAN DIEGO • SAN FRANCISCO • SINGAPORE • SYDNEY • TOKYO

Elsevier
32 Jamestown Road, London NW1 7BY
225 Wyman Street, Waltham, MA 02451, USA

Notices
Knowledge and best practice in this field are constantly changing. As new research and experience broaden our understanding, changes in research methods, professional practices, or medical treatment may become necessary.

Practitioners and researchers must always rely on their own experience and knowledge in evaluating and using any information, methods, compounds, or experiments described herein. In using such information or methods they should be mindful of their own safety and the safety of others, including parties for whom they have a professional responsibility.

To the fullest extent of the law, neither the Publisher nor the authors, contributors, or editors, assume any liability for any injury and/or damage to persons or property as a matter of products liability, negligence or otherwise, or from any use or operation of any methods, products, instructions, or ideas contained in the material herein.

British Library Cataloguing-in-Publication Data
A catalogue record for this book is available from the British Library

Library of Congress Cataloging-in-Publication Data
A catalog record for this book is available from the Library of Congress

ISBN: 978-0-323-28266-6

This book has been manufactured using Print On Demand technology. Each copy is produced to order and is limited to black ink. The online version of this book will show color figures where appropriate.

Transferred to Digital Printing in 2016

Contents

Preface

Under physical laws, the Anharmonicity becomes the Harmony...

This book is devoted to a new approach to the physical perturbation theory of systems according to quantum mechanics, in which anharmonic phenomena of vibrations of varied nature play an important role. Its main purpose is to yield a concise and precise presentation of the method arising from the evolution of the theory from traditional perturbation algebra to the more powerful methods of particular practice in physical problems. We introduce the formalism of polynomials of quantum numbers; its details are best revealed in various applications of the theory to solve concrete physical problems. In particular, one fruitful application of the polynomial formalism is in calculating the observable frequencies and intensities of lines in high-resolution spectra of molecules and their aggregates. The necessary factual information from the literature on quantum physics that precedes the material of the main content is collected in the first and introductory part of the book. The great textbooks on quantum mechanics by Bethe and Dirac, and likewise the third volume of the famous course of theoretical physics by Landau and Lifshitz, rendered invaluable assistance in the composition of this part.

The author is deeply grateful to colleagues Drs J.F. Ogilvie and A.A. Vigasin for valuable support during the preparation of the manuscript. I hope that this book becomes useful to a wide audience of readers, both theorists and experimenters, who specialize in the field of applied quantum mechanics.

Konstantin V. Kazakov
Irkutsk, Russian Federation
March 2012

1 The Laws of Quantum Mechanics

Introduction

The statistical character of physical states and Hamilton's formalism of classical mechanics form a fundament of quantum theory. We begin our discussion from the description of states with an example of the phenomenon of the polarization of light. Let the light waves, together with separate photons of this light beam, possess a particular polarization. We pass such a beam through a plate of tourmaline; on passing this crystal through unpolarized light, on the back of the plate we discover waves having the electric-field vector parallel to the optic axis of the crystal. If the electric-field vector in our beam is perpendicular to the optic axis, then as a result the entire absorption becomes observable. If the light is polarized at angle α to the axis, only a fraction equal to $\cos^2\alpha$ from the initial beam passes through the crystal. From the point of view of classical optics, these facts are trivial. The question arises, however, in the case of separate photons, whether each photon is polarized at angle α to the axis. The answer is simple: if we pass photons one by one from our beam, we discover that one photon is entirely transmitted, whereas another is entirely absorbed; the probability of observing a particular photon from the beam is equal to $\cos^2\alpha$, and the probability of its absorption is $\sin^2\alpha$. As a principle of quantum theory, one might thus apply the next device. Each photon can be represented in a state with polarization that is parallel to the axis or perpendicular to the axis. A particular superposition of these states produces the necessary state for the beam with polarization. In the result of an experiment, photons jump from an uncertain state to a state with a concrete polarization – those that pass and those that become absorbed.

The same condition occurs for the interference of photons. If an initial beam becomes split into two components, each photon with a particular weight enters partly into each component beam. As we have observed, however, that a particular photon is entirely in one component, it is at once precluded from being in the other component beam. A priori we may characterize a physical system with states of a particular number that have a statistical character. Quantum mechanics requires that each photon interferes only with itself during the interference of the two components. An electromagnetic wave and a photon are two descriptions of light. The same condition, as we see further, applies for physical particles with which one

Quantum Theory of Anharmonic Effects in Molecules. DOI: http://dx.doi.org/10.1016/B978-0-12-397912-4.00001-5

might also associate individual wave fields. In this sense, the individuality emphasizes the stability of all material – electrons, protons and so on.

Let us generalize the facts above. What should we understand about the state of the system – a motion, a rest, an interaction? These concepts exist in classical mechanics. Something similar holds in quantum mechanics, but it is less determinate. What is the meaning therein? If the system is presumably in one state, we must consider that it is partly in another state, so that its real state represents the superposition of all possible states that have non-zero probabilities. As a classical analogue of the expression of this principle, one might apply a wave packet, for which a complicated wave motion is resolvable into Fourier components; through this analogy, quantum mechanics is generally called wave mechanics. As a result, this principal idea yields a new theory – a theory of probabilities or amplitudes of physical states.

For states in quantum mechanics, as far as practicable, we use Dirac's notation. In this case, to each state we ascribe a ket vector $|\cdots\rangle$, inside of which might appear letters, words, numbers and other symbols. Keep in mind that in classical mechanics a vector is also applied to describe motion, but it is Euclidian there, whereas here Hilbert's type prevails. Vectors $|A_1\rangle$, $|A_2\rangle, \ldots$ that belong to a Hilbert space might be added together and might be multiplied by arbitrary complex numbers c_1, $c_2, \ldots$, as a result of which we obtain another vector

$$|A\rangle = c_1|A_1\rangle + c_2|A_2\rangle + \cdots.$$

This vector, which is expressible in a form of linear combination of others, is linearly dependent on them. Like a Euclidean space, the systems of linearly independent vectors are therefore of special interest. Each physical state of interest is expressible as an expansion in terms of these system vectors. Conversely, any such state might describe a concrete state of a physical system. It is important that a procedure of multiplying the vector by the number gives no new state; for instance, $|A\rangle$ and $-|A\rangle$ describe one and the same state. The principle of superposition in quantum mechanics has an important significance; considering the concrete physical problems, we generally appeal to this postulate.

Let us now consider Hamilton's formalism, which we will review briefly with regard to methods of classical mechanics. It is remarkable that the equations of the *old* theory can be borrowed with a somewhat altered meaning to construct the *new* mechanics. Lagrange's function of a mechanical system represents a function of generalized coordinates q_i, their temporal derivatives $\dot{q}_i$ (generalized velocities) and time t:

$$\Im = \Im(q_i, \dot{q}_i, t).$$

By definition, the momentum is $p_i = \partial\Im/\partial\dot{q}_i$ and the force is $F_i = \partial\Im/\partial q_i$. The energy of the system equals

$$H = \sum_i p_i\dot{q}_i - \Im.$$

Lagrange's function $\Im$ is such that integral $\int_{t_1}^{t_2} \Im \, \mathrm{d}t$ has a minimum; this condition leads to the Euler–Lagrange equation

$$\frac{\mathrm{d}}{\mathrm{d}t}\left(\frac{\partial \Im}{\partial \dot{q}_i}\right) - \frac{\partial \Im}{\partial q_i} = 0\left(\frac{\mathrm{d}p_i}{\mathrm{d}t} - F_i = 0\right)$$

if $\partial \Im / \partial q_i = 0$, p_i is a constant of motion and q_i is a cyclic coordinate.

There exists, however, an alternative method to describe a mechanical system that employs the language of coordinates and momenta. To convert to variables q_i and p_i, we apply a Legendre transformation:

$$\begin{aligned}\mathrm{d}H = \mathrm{d}\left(\textstyle\sum_i p_i \dot{q}_i - \Im\right) &= -\partial \Im + \textstyle\sum_i \left(-\frac{\partial \Im}{\partial q_i} \mathrm{d}q_i - \frac{\partial \Im}{\partial \dot{q}_i} \mathrm{d}\dot{q}_i + p_i \, \mathrm{d}\dot{q}_i + \dot{q}_i \, \mathrm{d}p_i \right) \\ &= -\partial \Im - \textstyle\sum_i \dot{p}_i \, \mathrm{d}q_i + \textstyle\sum_i \dot{q}_i \, \mathrm{d}p_i.\end{aligned}$$

Consequently,

$$\frac{\partial H}{\partial p_i} = \dot{q}_i, \qquad \frac{\partial H}{\partial q_i} = -\frac{\partial \Im}{\partial q_i} = -F_i = -\dot{p}_i, \qquad \frac{\partial H}{\partial t} = -\frac{\partial \Im}{\partial t}.$$

Here, H is Hamilton's function; this description is called a Hamiltonian formalism. One sees that this method possesses great symmetry. Moreover, it is convenient that H represents the total energy of the system. For instance, for interacting particles, the energy comprises kinetic and potential contributions:

$$H = \sum_i \frac{p_i^2}{2m_i} + V(q_1, q_2, \ldots),$$

in which m_i is the mass of particle i and V is the potential energy of interaction of the particles. In this case, Lagrange's function has a form

$$\Im = \sum_i \frac{m_i \dot{q}_i^2}{2} - V(q_1, q_2, \ldots).$$

In Hamilton's formalism, physical quantity f is represented as a function of the coordinates, momenta and time: $f(q_i, p_i, t)$. Its total derivative with respect to time has a form

$$\frac{\mathrm{d}f}{\mathrm{d}t} = \frac{\partial f}{\partial t} + \sum_i \frac{\partial f}{\partial q_i}\frac{\partial q_i}{\partial t} + \sum_i \frac{\partial f}{\partial p_i}\frac{\partial p_i}{\partial t} = \frac{\partial f}{\partial t} + \sum_i \frac{\partial f}{\partial q_i}\frac{\partial H}{\partial p_i} - \sum_i \frac{\partial f}{\partial p_i}\frac{\partial H}{\partial q_i} \equiv \frac{\partial f}{\partial t} + \{f, H\}.$$

Here,

$$\{f,H\} = \sum_i \left(\frac{\partial f}{\partial q_i} \frac{\partial H}{\partial p_i} - \frac{\partial f}{\partial p_i} \frac{\partial H}{\partial q_i} \right)$$

is a Poisson bracket. For instance, $\dot{p}_i = \{p_i, H\}$ and $\dot{q}_i = \{q_i, H\}$. Poisson brackets play an important role not only in classical mechanics but also in quantum theory; they therefore deserve special attention.

As an example, we consider the Hamiltonian of a particle in an external electromagnetic field, which is determined by vector potential **A** and scalar potential U. The energy of this particle with charge e' and velocity **v** in such a field is given with this expression

$$e'U - \frac{e'}{c}\mathbf{A} \cdot \mathbf{v},$$

in which c is the speed of light. For Lagrange's function, we thus have

$$\Im = \frac{m\mathbf{v}^2}{2} - e'U + \frac{e'}{c}\mathbf{A} \cdot \mathbf{v}.$$

The momentum is

$$\mathbf{p} = \frac{\partial \Im}{\partial \mathbf{v}} = m\mathbf{v} + \frac{e'}{c}\mathbf{A}.$$

By definition, we write expression for Hamiltonian H:

$$\mathbf{p} \cdot \mathbf{v} - \Im = m\mathbf{v}^2 + \frac{e'}{c}\mathbf{A} \cdot \mathbf{v} - \Im = \frac{m\mathbf{v}^2}{2} + e'U.$$

However, $\mathbf{v} \rightarrow (\mathbf{p} - e'\mathbf{A}/c)/m$, so that finally

$$H = \frac{1}{2m}\left(\mathbf{p} - \frac{e'}{c}\mathbf{A}\right)^2 + e'U.$$

One sees that, to proceed from the Hamiltonian of the freely moving particle to the Hamiltonian describing the motion in the external field, one must perform a replacement $\mathbf{p} \rightarrow \mathbf{p} - e'\mathbf{A}/c$ and add a trivial static energy $e'U$. Elsewhere in what follows, classical mechanics in Hamilton's form becomes the initial point of our research and prompts the correct form of initial equations.

Observables and Variables

To describe states in quantum mechanics, we introduced the concept of a vector. This definition is highly abstract; one must understand how to work with it. An experiment produces numerical values of physical quantities, which are involved in classical theory. This concept fails to be usable in quantum mechanics. We cannot directly operate with conventional numbers, in brief, c-numbers, or emphasize their triviality. The language of quantum mechanics involves q-numbers. If the coordinate and momentum are c-numbers in classical physics, in quantum physics they become q-numbers. The new numbers represent a new set of dynamical variables, namely those that we must treat. These variables are just determined in a space of abstract vectors – vectors of a Hilbert space. Through the action of q-numbers, such as some operation involving quantity O of q-type on some vector $|\varphi\rangle$, we obtain another vector $|\psi\rangle$. One might state that, in the simplest case, q-numbers are convenient operators, and questions of quantum mechanics consist of extracting observable c-numbers from a theory of dynamical variables of q-type.

Let us discuss the mathematical basis of quantum mechanics.

In a separable Hilbert space, vectors $|\varphi\rangle, |\psi\rangle, \ldots$ form a countably infinite sequence. For any pair of $|\varphi\rangle$ and $|\psi\rangle$, the sum $|\varphi\rangle + |\psi\rangle$ is determined, which is also a vector and possesses the properties commutativity and associativity:

$$|\varphi\rangle + |\psi\rangle = |\psi\rangle + |\varphi\rangle \text{ and } |\varphi\rangle + (|\psi\rangle + |\chi\rangle) = (|\varphi\rangle + |\psi\rangle) + |\chi\rangle.$$

The multiplication of vector $|\varphi\rangle$ by complex number c is defined; product $c|\varphi\rangle$ represents the vector and has the property distributivity:

$$c(|\varphi\rangle + |\psi\rangle) = c|\varphi\rangle + c|\psi\rangle, \quad (c+d)|\varphi\rangle = c|\varphi\rangle + d|\varphi\rangle.$$

Moreover,

$$1 \cdot |\varphi\rangle = |\varphi\rangle \text{ and } 0 \cdot |\varphi\rangle = 0.$$

Any two vectors $|\varphi\rangle$ and $|\psi\rangle$ possess a scalar product

$$\langle\varphi|\psi\rangle,$$

in which $\langle\varphi|$ is a so-called bra vector that is the complex conjugate of $|\varphi\rangle$. Obviously, $\langle\varphi|\varphi\rangle \geq 0$, and $\langle\varphi|\varphi\rangle = 0$ only in the case when $|\varphi\rangle = 0$. If $|\psi\rangle$ represents the sum $|\theta\rangle + |\chi\rangle$,

$$\langle\varphi|\psi\rangle = \langle\varphi|\theta\rangle + \langle\varphi|\chi\rangle;$$

if $|\psi\rangle$ equals vector $|\chi\rangle$ that is multiplied by number c,

$$\langle\varphi|\psi\rangle = c\langle\varphi|\chi\rangle.$$

Finally, $\langle\varphi|\psi\rangle^* = \langle\psi|\varphi\rangle$.

For the vectors in a Hilbert space, these properties are general. As an example, we consider a case in which as vectors $|\varphi\rangle, |\psi\rangle, \ldots$ we have ordinary functions

$\varphi(x)$, $\psi(x), \ldots$, which are determined in manifold G. It is generally convenient to apply this representation to solve concrete problems of quantum mechanics. The properties of vectors, in this case, are performed in such a manner:

$\varphi(x) + \psi(x)$ – sum $\varphi(x)$ and $\psi(x)$ in G;

$c\varphi(x)$ – multiplication by a number;

$\langle\varphi|\psi\rangle = \int_G \varphi^*(x)\psi(x)\mathrm{d}x$ – scalar product.

For each vector, one might introduce the definition of length or norm that, in the sense of a number, equals $||\varphi|| = \sqrt{\langle\varphi|\varphi\rangle}$. If $||\varphi|| = 1$, vector $|\varphi\rangle$ is normalized. If the scalar product of two vectors $|\varphi\rangle$ and $|\psi\rangle$ equals zero, i.e. $\langle\varphi|\psi\rangle = 0$, these vectors are orthogonal. The set of orthonormal vectors might represent a complete basis. Considering the physical principle of superposition, we have already mentioned the necessity of the condition completeness for states. A sequence of vectors $|\varphi_i\rangle$ is mathematically complete if any vector $|\Phi\rangle$ in a certain space is expressible in a form of linear combination:

$$|\Phi\rangle = \sum_i c_i|\varphi_i\rangle.$$

Vectors $|\varphi_i\rangle$ are linearly independent only in the case in which there is no relation of type $\sum_i c_i|\varphi_i\rangle = 0$, eliminating the case $c_i = 0$.

With the aid of a convenient operator, one might convert one vector into another. For instance, square root $\sqrt{\ldots}$ and differentiation $\mathrm{d}(\ldots)/\mathrm{d}x$ are simple operators. Not all operators, however, represent a physical interest, and a mathematical operation should not be associated with a dynamical variable; only a few of them are applicable in physics. We imply here linear operators O that play an exceptional role in quantum mechanics. Quantity O implies some rule according to which a vector, e.g. $|\varphi\rangle$, transforms into $|\psi\rangle$. Linearity means that

$$O(a|\varphi\rangle + b|\psi\rangle) = aO|\varphi\rangle + bO|\psi\rangle,$$

in which a and b are c-numbers. As simple examples of linear operations, one might undertake multiplication by an arbitrary coordinate function – $F(x)\varphi(x)$, or differentiation – $\mathrm{d}^n\varphi(x)/\mathrm{d}x^n$. Exponentiation of a vector to some power as $\varphi^n(x)$ is, however, not a linear operation.

Let us enumerate the general properties of linear operators. For any pair of operators A and B, sum $A + B$ is defined:

$$(A + B)|\varphi\rangle = A|\varphi\rangle + B|\varphi\rangle;$$

such a sum possesses properties commutativity and associativity:

$$(A + B)|\varphi\rangle = (B + A)|\varphi\rangle \text{ and } A|\varphi\rangle + (B + C)|\varphi\rangle = (A + B)|\varphi\rangle + C|\varphi\rangle.$$

The multiplication of operator A by complex number c is determined:

$$(c \cdot A)|\varphi\rangle = c(A|\varphi\rangle).$$

There is determined the product of operators $A \cdot B$ with properties distributivity

$$A(B + C)|\varphi\rangle = AB|\varphi\rangle + AC|\varphi\rangle,$$

associativity

$$(AB)|\varphi\rangle = A(B|\varphi\rangle)$$

and, generally, non-commutativity

$$AB|\varphi\rangle \neq BA|\varphi\rangle.$$

The principal role belongs to the commutator of two operators

$$[A, B] = AB - BA = -[B, A];$$

obviously, $[A, A] = 0$. For instance, if vector $|\varphi\rangle$ is function φ of variable x,

$$\left[\mathrm{d}/\mathrm{d}x, x\right]\varphi(x) = \frac{\mathrm{d}}{\mathrm{d}x}(x\varphi) - x\frac{\mathrm{d}\varphi}{\mathrm{d}x} = \varphi(x),$$

where from

$$[\mathrm{d}/\mathrm{d}x, x] = 1.$$

If equation $A|\varphi\rangle = |\psi\rangle$ is solvable with regard to $|\varphi\rangle$, such that there exists a relation of type $|\varphi\rangle = B|\psi\rangle$, operator B, which is equal, by definition, to A^{-1}, is called reciprocal to A. So

$$(A^{-1}A)|\varphi\rangle = A^{-1}(A|\varphi\rangle) = A^{-1}|\psi\rangle = |\varphi\rangle,$$

i.e. $A^{-1}A = 1$, and also $AA^{-1} = 1$, hence $[A, A^{-1}] = 0$.

The product of the same operators yields a concept of power n of an operator:

$$A^n|\varphi\rangle = A(A(A\ldots(A|\varphi\rangle)\ldots));$$

in particular, if $A = \mathrm{d}/\mathrm{d}x$, then $A^n = \mathrm{d}^n/\mathrm{d}x^n$. With the aid of an exponentiation operation, one might determine function f of an operator:

$$f(A) = \sum_{i=0}^{\infty}\frac{f^{(i)}(0)}{i!}A^i.$$

For instance, we consider function

$$e^{\alpha(d/dx)} = \sum_{n=0}^{\infty} \frac{\alpha^n}{n!} \frac{d^n}{dx^n};$$

on acting on $\varphi(x)$, we have

$$e^{\alpha(d/dx)}\, \varphi(x) = \sum_{n=0}^{\infty} \frac{\alpha^n}{n!} \frac{d^n \varphi}{dx^n} = \varphi(x+\alpha).$$

Operator $e^{\alpha(d/dx)}$ thus shifts the argument of function $\varphi(x)$ by quantity α.

Furthermore, if there exists an equation

$$A|\varphi\rangle = a|\varphi\rangle,$$

in which a is a c-number, quantities a represent eigenvalues of operator A and $|\varphi\rangle$ are its eigenfunctions. Let us draw an important conclusion. Suppose that A and B are commutative operators, then

$$B(A|\varphi\rangle) = B(a|\varphi\rangle) \text{ and } A(B|\varphi\rangle) = a(B|\varphi\rangle).$$

One sees that vector $B|\varphi\rangle$ is an eigenvector of operator A, and there must exist a relation of type

$$B|\varphi\rangle = b|\varphi\rangle,$$

in which b is a c-number. Thus, if $[A,B] = 0$, A and B have simultaneously a complete system of eigenvectors (eigenfunctions).

Linear operator A in some basis can be represented with a matrix. This condition is easy to understand if we suggest that we have a complete system of vectors $|\varphi_i\rangle$, and arbitrary vector $|\psi\rangle$ is expressible in a form of this expansion

$$|\psi\rangle = \sum_i c_i|\varphi_i\rangle,$$

in which c_i are coefficients. On action by operator A on $|\psi\rangle$, we have

$$A|\psi\rangle = \sum_i c_i A|\varphi_i\rangle,$$

where from

$$\langle\varphi_k|A|\psi\rangle = \sum_i c_i\langle\varphi_k|A|\varphi_i\rangle \equiv \sum_i c_i A_{ki}.$$

The complete set of matrix elements A_{ki} forms a matrix representing the linear operator. To coefficients c_i one might ascribe a physical meaning of amplitudes of the states; then $|c_i|^2$ is the probability of state $|\varphi_i\rangle$, and a sum of all probabilities equals unity:

$$\langle\psi|\psi\rangle = \sum_i |c_i|^2 = 1.$$

For instance, unit operator I is defined with equation $I|\varphi\rangle = |\varphi\rangle$ and might be represented with unit matrix δ_{ki}, in which $\delta_{kk} = 1$ and $\delta_{ki} = 0$ for $k \neq i$.

We consider an expression for diagonal matrix element,

$$\langle\psi|A|\psi\rangle = \sum_{ik} c_i^* c_k \langle\varphi_i|A|\varphi_k\rangle.$$

If $|\varphi_i\rangle$ and a_i are eigenvectors and eigenvalues of operator A,

$$\langle\psi|A|\psi\rangle = \sum_i |c_i|^2 a_i$$

that represents the mathematical expectation value of quantity A. In state $|\psi\rangle$, the expectation value of a dynamical variable (operator) is thus determined by the diagonal matrix element

$$\langle A\rangle = \langle\psi|A|\psi\rangle = \text{(in particular)} \sum_i |c_i|^2 a_i.$$

In quantum theory, the expectation values belong to a class of observable quantities.

According to a definition,

$$\langle\varphi|A|\psi\rangle = \langle\psi|A^+|\varphi\rangle^*,$$

to every linear operator A one might determine Hermitian conjugate operator A^+. One sees that

$$(A^+)^+ = A, (A+B)^+ = A^+ + B^+ \text{ and } (AB)^+ = B^+A^+.$$

If $A = A^+$, A is called the self-adjoint or Hermitian operator; in this case,

$$\langle\varphi|A|\psi\rangle = \langle\psi|A|\varphi\rangle^*.$$

Hermitian operators play an important role in quantum mechanics. For instance, we consider a dynamical variable that is described with operator A. Suppose furthermore that, in some state $|\varphi\rangle$, our variable equals a certain c-number a; then $A|\varphi\rangle = a|\varphi\rangle$, and, consequently, $\langle\varphi|A|\varphi\rangle = a$. If operator A is Hermitian,

$$\langle\varphi|A|\varphi\rangle = \langle\varphi|A^+|\varphi\rangle^* = \langle\varphi|A|\varphi\rangle^*,$$

thus, $a = a^*$, and the eigenvalues of A are real numbers. In physics, the dynamical variables in some arbitrary states must have only real values. The condition of hermitivity is, therefore, generally necessary to ascribe some operator to a physical quantity.

Another important consequence deserves attention. Let a and a' be eigenvalues of operator A with corresponding vectors $|\varphi\rangle$ and $|\varphi'\rangle$. Then

$$A|\varphi\rangle = a|\varphi\rangle \quad \text{and} \quad \langle\varphi'|A|\varphi\rangle = a\langle\varphi'|\varphi\rangle.$$

On the other side, if A is a Hermitian operator,

$$\langle\varphi'|A = a'\langle\varphi'| \quad \text{and} \quad \langle\varphi'|A|\varphi\rangle = a'\langle\varphi'|\varphi\rangle.$$

We see that

$$(a - a')\langle\varphi'|\varphi\rangle = 0.$$

Thus, if $a \neq a'$, $\langle\varphi'|\varphi\rangle = 0$; two eigenvectors of a Hermitian dynamical variable belonging to various eigenvalues are orthogonal.

The Conditions of Quantum Theory

Comparing dynamical variables with linear operators, we understand that one might scarcely succeed to preserve in their original form the equations of classical mechanics. Those operators generally fail to conform to the commutative conditions. Because of this obstacle, we cannot build quantum theory using only experimental relations for physical quantities: we must invoke additional relations on q-numbers – the quantum conditions. These conditions are generally expressible through the commutators of the particular variables. For every such pair, there exists a certain commutator. The determination of all necessary commutators is an indispensable condition a priori, without which it might be impossible to find a solution. There is no way to write all conditions in a unified manner; in some way, they are individual. One might, however, reveal some similarities with classical theory. It turns out that the properties of commutators closely resemble those of classical Poisson brackets.

We consider a Poisson bracket for variables A and B, which are functions of canonical coordinates q_i and momenta p_i,

$$\{A, B\} = \sum_i \left(\frac{\partial A}{\partial q_i}\frac{\partial B}{\partial p_i} - \frac{\partial A}{\partial p_i}\frac{\partial B}{\partial q_i} \right).$$

If one variable is constant number c,

$$\{A, c\} = 0.$$

If we exchange quantities A and B within the braces, the sign is automatically reversed:

$$\{A, B\} = -\{B, A\}.$$

If $A \to A + A'$ or $B \to B + B'$,

$$\{A + A', B\} = \{A, B\} + \{A', B\} \text{ and } \{A, B + B'\} = \{A, B\} + \{A, B'\}.$$

If $A \to AA'$ or $B \to BB'$,

$$\{AA', B\} = \{A, B\}A' + A\{A', B\} \text{ and } \{A, BB'\} = \{A, B\}B' + B\{A, B'\}.$$

These properties are simple and understandable, because in classical mechanics, through commutativity, the order of various dynamical variables has no principal significance. In quantum theory, commutativity generally has no place, and a Poisson bracket must be redefined.

Let quantum bracket

$$\{A, B\}$$

possess properties similar to those of a classical Poisson bracket, and the variables generally fail to conform to the law of commutative multiplication. We calculate $\{AA', BB'\}$; on the one side,

$$\begin{aligned}\{AA', BB'\} &= A\{A', BB'\} + \{A, BB'\}A' \\ &= AB\{A', B'\} + A\{A', B\}B' + \{A, B\}B'A' + B\{A, B'\}A',\end{aligned}$$

on the other,

$$\begin{aligned}\{AA', BB'\} &= B\{AA', B'\} + \{AA', B\}B' \\ &= BA\{A', B'\} + B\{A, B'\}A' + A\{A', B\}B' + \{A, B\}A'B'.\end{aligned}$$

Consequently,

$$(AB - BA)\{A', B'\} = \{A, B\}(A'B' - B'A').$$

Comparing the left and right sides of this obtained equality, we see that the commutator equals the Poisson bracket that is accurate within a constant coefficient. By definition, we have

$$\underset{\text{commutator}}{AB - BA} = \underset{\text{constant}}{i\hbar} \cdot \underset{\text{Poisson bracket}}{\{A, B\}}.$$

Constant $\hbar$, introduced by Dirac, is related trivially to the universal Planck constant h through relation $\hbar = h/2\pi$.

In quantum theory, the condition of non-commutativity of the dynamical variables yields absolutely another definition of a Poisson bracket. An imaginary unit, which is specially introduced, emphasizes that in the classical understanding of the dynamical variables, for instance, coordinates and momenta, there are no conditions of type

$$AB - BA = \mathrm{i}\hbar\{A, B\}.$$

These conditions appear in quantum mechanics. Each condition is a result of a classical Poisson bracket on the one side and the commutator divided by $\mathrm{i}\hbar$ on the other side. Thus,

$$\{A, B\} = \frac{1}{\mathrm{i}\hbar}[A, B],$$

which is entirely correct because the commutators are characterized by a set of properties similar to those of the Poisson brackets, in particular

$$[A, c] = 0, \quad [A + A', B] = [A, B] + [A', B]$$
$$\text{and } [AA', B] = [A, B]A' + A[A', B].$$

To proceed to quantum mechanics, we demand that principal relations between canonical coordinates q_i and momenta p_i preserve their form. We have

$$\{q_i, q_j\} = 0, \quad \{p_i, p_j\} = 0 \text{ and } \{q_i, p_j\} = \delta_{ij},$$

hence

$$q_i q_j - q_j q_i = 0, \quad p_i p_j - p_j p_i = 0$$

and

$$q_i p_j - p_j q_i = \mathrm{i}\hbar\delta_{ij}.$$

One might write the quantum conditions for other dynamical variables that represent the expansions in terms of conjugate coordinates and momenta. The quantum conditions give us a boundary between classical and quantum theories. If $\hbar$ tends to zero, quantity $AB - BA$ also becomes equal to zero. Neglecting small constant $\hbar$, we perform the limiting conversion from quantum mechanics to classical.

We consider the coordinate, or Schrödinger's, representation. In this case, coordinates q_i represent the pertinent variables, vectors of the states φ are functions of the coordinates and momenta p_i are some operators. Applying this quantum condition,

$$[q_i, p_i] = \mathrm{i}\hbar,$$

we must determine a form of p_i. We have

$$[q_i, p_i]\varphi(q) = \mathrm{i}\hbar\varphi(q),$$

in which q implies the complete set of quantities q_i. Dividing both sides of this equation by i$\hbar$, one obtains

$$[q_i, p_i/\mathrm{i}\hbar]\varphi(q) = \varphi(q).$$

The latter expression becomes satisfied if we assume $p_i/\mathrm{i}\hbar = -\partial/\partial q_i$ and recall that

$$[\partial/\partial q_i, q_i] = 1.$$

Momentum is thus a differential operator,

$$p_i = -\,\mathrm{i}\hbar\frac{\partial}{\partial q_i}.$$

Choosing, instead of q_i, Cartesian coordinates x, y and z, for instance, we have

$$p_x = -\,\mathrm{i}\hbar\frac{\partial}{\partial x}, \quad p_y = -\,\mathrm{i}\hbar\frac{\partial}{\partial y}, \quad p_z = -\,\mathrm{i}\hbar\frac{\partial}{\partial z};$$

if $\mathbf{r} = (x, y, z)$ and $\mathbf{p} = (p_x, p_y, p_z)$,

$$\mathbf{p} = -\,\mathrm{i}\hbar\frac{\partial}{\partial \mathbf{r}} = -\,\mathrm{i}\hbar\nabla.$$

Quantities p_x, p_y and p_z are commutative and together with **p** can thus be measured in one state. Momentum **p** and some arbitrary function $f(\mathbf{r})$ are not simultaneously measurable; in this case,

$$[f(\mathbf{r}), \mathbf{p}] = \mathrm{i}\hbar\ \nabla f(\mathbf{r}) \neq 0.$$

In the momentum representation, the pertinent variables are p_i, vectors φ depend functionally on momenta and coordinates q_i become operators. Repeating actions similar to those that we make in the coordinate representation above, we obtain an expression for operator q_i:

$$q_i = \mathrm{i}\hbar\frac{\partial}{\partial p_i}.$$

In quantum mechanics, one might thus maintain a proper symmetry between the canonical conjugate variables – coordinates and momenta.

One might indirectly confirm the correctness of this choice, e.g. for the operator of momentum, on considering the problem on eigenvalues of quantity p. For this purpose, we must solve this equation:

$$-\mathrm{i}\hbar\frac{\partial}{\partial q}\varphi(q) = p'\varphi(q),$$

in which p' are the sought eigenvalues of the momentum. As a result,

$$\varphi(q) = C\,\mathrm{e}^{\mathrm{i}kq},$$

in which $k = p'/\hbar$ and C is a constant of integration. We arrived at a conventional de Broglie wave describing the state of a freely moving particle with a particular momentum; as follows from the solution, the possible values of p' run from $-\infty$ to $+\infty$. It is important to note that functions $\varphi(q)$ are normalized not to unity, but to Dirac's delta function, i.e.

$$|C|^2 \int_{-\infty}^{+\infty} \mathrm{e}^{\mathrm{i}(k-k')q}\,\mathrm{d}q = |C|^2 2\pi\delta(k-k') = \delta(k-k').$$

Having assumed $C = 1/\sqrt{2\pi}$, we eventually obtain normalized eigenfunctions of the operator of momentum in a form $\varphi(q) = \mathrm{e}^{\mathrm{i}kq}/\sqrt{2\pi}$.

Analysing the above facts, we see that quantities q and p are not measurable in one state. Let us consider this question in detail. Suppose that we have some state $|\varphi\rangle$; the expectation values of q and p in this state are equal to

$$\langle q\rangle = \langle\varphi|q|\varphi\rangle \quad\text{and}\quad \langle p\rangle = \langle\varphi|p|\varphi\rangle,$$

and the corresponding dispersions are

$$(\Delta q)^2 = \langle\varphi|(q-\langle q\rangle)^2|\varphi\rangle \quad\text{and}\quad (\Delta p)^2 = \langle\varphi|(p-\langle p\rangle)^2|\varphi\rangle.$$

We introduce an auxiliary dynamical variable

$$A = (q - \langle q\rangle) + \mathrm{i}a(p - \langle p\rangle),$$

in which a is a real positive quantity, and consider matrix element $\langle\varphi||A|^2|\varphi\rangle$:

$$\langle\varphi|A^*A|\varphi\rangle = (\Delta q)^2 + a^2(\Delta p)^2 + \mathrm{i}a\langle\varphi|[q,p]|\varphi\rangle.$$

As $[q,p] = \mathrm{i}\hbar$,

$$\langle\varphi|A^*A|\varphi\rangle = (\Delta q)^2 + a^2(\Delta p)^2 - a\hbar.$$

Taking into account that $\langle\varphi||A|^2|\varphi\rangle \geq 0$, we obtain

$$(\Delta q)^2 + a^2(\Delta p)^2 - a\hbar \geq 0,$$

where from

$$a^{-1}(\Delta q)^2 + a(\Delta p)^2 \geq \hbar.$$

Each quantity, $(\Delta q)^2$ and $(\Delta p)^2$, is obviously greater than zero; assuming a as some parameter, we determine a minimum of function

$$a^{-1}(\Delta q)^2 + a(\Delta p)^2$$

from the condition that its first derivative equals zero:

$$-a^{-2}(\Delta q)^2 + (\Delta p)^2 = 0.$$

We have $a = \Delta q/\Delta p$ and

$$2\,\Delta q\,\Delta p$$

is the sought minimum. Our inequality, in this case, leads to a form

$$\Delta q\,\Delta p \geq \frac{\hbar}{2}.$$

This result is the famous Heisenberg principle of indeterminacy demonstrating that the uncertainty in momentum increases as the uncertainty in coordinate decreases, and vice versa. There is thus no state for which all values either of coordinate at a particular momentum or of momentum at a particular coordinate are equally probable. A physical explanation is that, in the case of coordinate, there exists a limitation on the size of the system and, in the case of momentum, there exists a limitation on energy. One readily observes that the classical limit remains valid: as $\hbar \to 0$ we obtain the complete certainty of both momentum and coordinate.

Angular Momentum

In quantum mechanics, angular momentum that has a dimension the same as that of a Planck constant plays an important role, just as in classical physics. Like the total energy, angular momentum **L** of an isolated system is a constant of the motion. Through the isotropy of space, this law is concerned with the symmetry with respect to rotations of a coordinate system. For a particle moving in a field of central forces, the angular momentum about the origin is conserved. For a particle in a field with axial symmetry, the projection of quantity **L** along the symmetry axis is invariant. The law of conservation of angular momentum is generally not fulfilled.

Let $\mathbf{r} = (x, y, z)$ be the radius vector of a particle, and $\mathbf{p} = (p_x, p_y, p_z)$ be its momentum; with the aid of the vector product of **r** and **p**, we introduce $\mathbf{L} = (L_x, L_y, L_z)$:

$$\mathbf{L} = \mathbf{r} \times \mathbf{p} = \begin{cases} yp_z - zp_y, \\ zp_x - xp_z, \\ xp_y - yp_x. \end{cases}$$

This definition is correct because variable y commutes with p_z, z with p_y, x with p_z and so on; there is thus no need to make concrete the order of various factors. As quantities **r** and **p** fail generally to commute with each other, **L** commutes with neither **r** nor **p**. We demonstrate this fact through a direct calculation of the commutation relations. We have

$$[L_x, x] = [yp_z - zp_y, x] = 0,$$
$$[L_x, y] = [yp_z - zp_y, y] = -z[p_y, y] = i\hbar z,$$
$$[L_x, z] = [yp_z - zp_y, z] = y[p_z, z] = -i\hbar y;$$

analogously

$$[L_x, p_x] = [yp_z - zp_y, p_x] = 0,$$
$$[L_x, p_y] = [yp_z - zp_y, p_y] = [y, p_y]p_z = i\hbar p_z,$$
$$[L_x, p_z] = [yp_z - zp_y, p_z] = -[z, p_z]p_y = -i\hbar p_y.$$

Other relations are obtainable through a cyclic permutation of x, y and z; for instance,

$$[L_y, z] = i\hbar x \rightarrow [L_z, x] = i\hbar y \rightarrow [L_x, y] = i\hbar z.$$

The commutation relations for **L** and **r** are hence exactly analogous to those for **L** and **p**. If $\mathbf{a} = (a_x, a_y, a_z)$ is **r** or **p**,

$$[L_x, a_x] = 0, \quad [L_x, a_y] = i\hbar a_z, \quad [L_x, a_z] = -i\hbar a_y, \ldots.$$

Let $\mathbf{b} = (b_x, b_y, b_z)$ also be **r** or **p**, then

$$[L_x, \mathbf{a} \cdot \mathbf{b}] = [L_x, a_x b_x + a_y b_y + a_z b_z]$$
$$= a_y[L_x, b_y] + [L_x, a_y]b_y + a_z[L_x, b_z] + [L_x, a_z]b_z = 0;$$

accordingly,

$$[L_y, \mathbf{a} \cdot \mathbf{b}] = 0 \text{ and } [L_z, \mathbf{a} \cdot \mathbf{b}] = 0.$$

Any scalar consisting of **a** and **b** thus commutes with **L**:

$$[L_i, \mathbf{r}^2] = 0, \quad [L_i, \mathbf{p}^2] = 0, \quad [L_i, \mathbf{r} \cdot \mathbf{p}] = 0, \ldots,$$

in which i denotes x or y or z.

We calculate the commutation relations for the components of angular momentum **L**:

$$[L_x, L_y] = [L_x, zp_x - xp_z] = [L_x, z]p_x - x[L_x, p_z] = i\hbar(xp_y - yp_x) = i\hbar L_z;$$

through cyclic permutations, we obtain other commutators

$$[L_y, L_z] = i\hbar L_x \text{ and } [L_z, L_x] = i\hbar L_y,$$

that are compactly expressible in a vector form

$$\mathbf{L} \times \mathbf{L} = i\hbar \mathbf{L}.$$

This formula is not quite absurd; one should bear in mind that components L_x, L_y and L_z fail to commute with each other.

Similar commutation relations are derivable for the case of the total angular momentum of a system of several particles. Let $\mathbf{L}_s$ be the angular momentum of particle s, then

$$\mathbf{L}_s \times \mathbf{L}_s = i\hbar \mathbf{L}_s$$

and

$$\mathbf{L}_s \times \mathbf{L}_j + \mathbf{L}_j \times \mathbf{L}_s = 0, \quad s \neq j.$$

If the total angular momentum equals $\mathbf{L} = \sum_s \mathbf{L}_s$,

$$\mathbf{L} \times \mathbf{L} = \sum_{s,j} \mathbf{L}_s \times \mathbf{L}_j = \sum_s \mathbf{L}_s \times \mathbf{L}_s + \sum_{s<j} (\mathbf{L}_s \times \mathbf{L}_j + \mathbf{L}_j \times \mathbf{L}_s) = i\hbar \sum_s \mathbf{L}_s = i\hbar \mathbf{L},$$

for which the proof was required.

We consider the squared angular momentum

$$\mathbf{L}^2 = L_x^2 + L_y^2 + L_z^2$$

and calculate commutator $[L_i, \mathbf{L}^2]$, in which index i denotes values x or y or z. We have

$$\begin{gathered}
[L_x, L_x^2] = 0, \\
[L_x, L_y^2] = L_y[L_x, L_y] + [L_x, L_y]L_y = i\hbar(L_yL_z + L_zL_y), \\
[L_x, L_z^2] = L_z[L_x, L_z] + [L_x, L_z]L_z = -i\hbar(L_zL_y + L_yL_z).
\end{gathered}$$

On summing these equalities, one finds

$$[L_x, \mathbf{L}^2] = 0.$$

In an analogous manner,

$$[L_y, \mathbf{L}^2] = 0 \text{ and } [L_z, \mathbf{L}^2] = 0.$$

The squared angular momentum, commuting with each component of vector **L**, might thus be simultaneously measured with one projection L_x or L_y or L_z. The projections of **L** fail to be commutative quantities with each other, and they are therefore not measurable in one state.

If projection L_z is defined, instead of indeterminate quantities L_x and L_y, it is convenient to choose another pair of operators

$$L_+ = L_x + \mathrm{i}L_y \text{ and } L_- = L_x - \mathrm{i}L_y.$$

One accordingly performs the next relations:

$$\begin{aligned}
[L_+, L_-] &= -\mathrm{i}[L_x, L_y] + \mathrm{i}[L_y, L_x] = 2\hbar L_z,\\
[L_z, L_+] &= [L_z, L_x] + \mathrm{i}[L_z, L_y] = \hbar L_+,\\
[L_z, L_-] &= [L_z, L_x] - \mathrm{i}[L_z, L_y] = -\hbar L_-,\\
\mathbf{L}^2 &= L_- L_+ + \hbar L_z + L_z^2 = L_+ L_- - \hbar L_z + L_z^2;
\end{aligned}$$

plus the well-known expressions for differential operators of angular momentum in spherical coordinates r, θ and ϕ:

$$L_\pm = \hbar\, \mathrm{e}^{\pm \mathrm{i}\phi}\left(\pm\frac{\partial}{\partial\theta} + \mathrm{i}ctg\theta\frac{\partial}{\partial\phi}\right), \quad L_z = -\mathrm{i}\hbar\frac{\partial}{\partial\phi},$$

$$\mathbf{L}^2 = -\hbar^2\left[\frac{1}{\sin^2\theta}\frac{\partial^2}{\partial\phi^2} + \frac{1}{\sin\theta}\frac{\partial}{\partial\theta}\left(\sin\theta\frac{\partial}{\partial\theta}\right)\right] \equiv -\hbar^2\nabla^2_{\theta\phi},$$

in which $\nabla^2_{\theta\phi}$ is an angular part of the Laplace operator.

We calculate eigenvalues of operators L_z and $\mathbf{L}^2$; in this representation, L_x and L_y have indeterminate values. Let φ be eigenvectors and L_z' be eigenvalues of L_z, then

$$-\mathrm{i}\hbar\frac{\partial}{\partial\phi}\varphi(\phi) = L_z'\,\varphi(\phi).$$

This equation is readily integrated; as a result,

$$\varphi(\phi) = \frac{1}{\sqrt{2\pi}}C(r,\theta)\mathrm{e}^{\mathrm{i}L_z'\phi/\hbar}.$$

Function φ must be periodic in ϕ; the eigenvalues of projection L_z are consequently integral multiples of $\hbar$:

$$L_z' = \hbar k, \quad k = 0, \pm 1, \pm 2, \pm 3, \ldots.$$

Here, $C(r,\theta)$ is a constant of integration; factor $1/\sqrt{2\pi}$ appears through a normalization condition

$$\frac{1}{2\pi}\int_0^{2\pi} e^{i(k'-k)\phi}\,d\phi = \delta_{k'k}.$$

If instead of L_z we choose, for instance, L_x, we arrive at the same result, but just for the x-component of the angular momentum; in this representation, projections L_z and L_y then have no determinate value. An exception to this rule is the case

$$L_x = L_y = L_z = 0;$$

then $\mathbf{L}^2 = 0$ and all projections of $\mathbf{L}$ are simultaneously measurable.

We proceed to calculate eigenvalues of the squared angular momentum. As $L_zL_+ = L_+L_z + \hbar L_+$, we have

$$L_zL_+|\varphi_k\rangle = \hbar(k+1)L_+|\varphi_k\rangle,$$

in which we took $L_z|\varphi_k\rangle = \hbar k|\varphi_k\rangle$ into account. Vector $L_+|\varphi_k\rangle$ is consequently the eigenvector of projection L_z belonging to eigenvalue $\hbar(k+1)$, that is accurate within a constant coefficient. Assume

$$|\varphi_{k+1}\rangle \sim L_+|\varphi_k\rangle.$$

In an analogous manner, applying commutator $[L_z,L_-] = -\hbar L_-$, one might obtain that

$$|\varphi_{k-1}\rangle \sim L_-|\varphi_k\rangle.$$

Thus, L_+ is the operator that increases the value of k by unity and L_- is the operator that decreases k by unity.

We apply the non-negativity of expression

$$\mathbf{L}^2 - L_z^2 = L_x^2 + L_y^2.$$

As

$$\mathbf{L}^2 - L_z^2$$

possesses only positive eigenvalues, there must exist an upper limit for L_z'; we denote it as $\hbar\ell$, in which ℓ is a positive integer. The states with $k > \ell$, by definition, do not exist; one must, therefore, satisfy the equation $L_+|\varphi_\ell\rangle = 0$. On acting on this equality with the lowering operator on the left, one obtains

$$L_-L_+|\varphi_\ell\rangle = (\mathbf{L}^2 - L_z^2 - \hbar L_z)|\varphi_\ell\rangle = 0.$$

Generally $|\varphi_\ell\rangle \neq 0$; denoting the eigenvalue of $\mathbf{L}^2$ as Λ, we have hence

$$\Lambda - \hbar^2\ell^2 - \hbar^2\ell = 0,$$

where from

$$\Lambda = \hbar^2\ell(\ell + 1).$$

Moreover, one should note these useful relations

$$L_\pm|\ell k\rangle = \hbar\sqrt{(\ell \mp k)(\ell \pm k + 1)}|\ell, k \pm 1\rangle,$$

which we implicitly applied and which are worthy of proof. We act in turn by raising and lowering operators on vector $|\varphi_k\rangle$, which is equal, by definition, to $|\ell k\rangle$; as a result,

$$\begin{aligned} L_-(L_+|\ell k\rangle) &= \hbar^2(\ell - k)(\ell + k + 1)|\ell k\rangle = (\hbar^2\ell(\ell + 1) - \hbar^2k^2 - \hbar^2k)|\ell k\rangle \\ &= (\mathbf{L}^2 - L_z^2 - \hbar L_z)|\ell k\rangle. \end{aligned}$$

As $L_-L_+ = \mathbf{L}^2 - L_z^2 - \hbar L_z$, the above relations become proved.

Thus,

$$\mathbf{L}^2|\ell k\rangle = \hbar^2\ell(\ell + 1)|\ell k\rangle, \quad \ell = 0,\ 1,\ 2, \ldots,$$

and

$$L_z|\ell k\rangle = \hbar k|\ell k\rangle, \quad k = 0,\ \pm 1, \ldots,\ \pm\ell.$$

Quantum number ℓ defines the squared angular momentum and might be equal to some non-negative integer. Quantities of projection $\mathbf{L}$ along a selected direction are integral multiples of constant $\hbar$. For each ℓ, quantity $L_z'/\hbar$ runs over all negative and positive integers from $-\ell$ to $+\ell$. As a result, the state with a particular and non-zero number ℓ becomes degenerate; the degeneracy numbers $2\ell + 1$; that many functions hence belong to eigenvalue $\hbar^2\ell(\ell + 1)$. Eigenfunctions $|\ell k\rangle$ satisfy the Laplace equation

$$[\nabla^2_{\theta\phi} + \ell(\ell + 1)]|\ell k\rangle = 0$$

and are represented as spherical harmonics,

$$Y_{\ell k}(\theta, \phi) = N(\ell, k)P_\ell^{|k|}(\cos\theta)\mathrm{e}^{\mathrm{i}k\phi},$$

in which $P_\ell^{|k|}(\cos\theta)$ are associated Legendre polynomials and $N(\ell,k)$ are normalized coefficients. Quantities $Y_{\ell k}(\theta,\phi)$ are orthonormal expansions in terms of $\sin\theta$, $\cos\theta$ and $\mathrm{e}^{\mathrm{i}\phi}$:

$$Y_{00} = \frac{1}{\sqrt{4\pi}}, \quad Y_{10} = \mathrm{i}\sqrt{\frac{3}{4\pi}}\cos\theta, \quad Y_{1,\pm 1} = \mp\mathrm{i}\sqrt{\frac{3}{8\pi}}\mathrm{e}^{\pm\mathrm{i}\phi}\sin\theta$$

and so on.

The Principal Equations

Up to this point, we consider the state vectors and dynamical variables with no regard to their temporal evolution. How can we trace the temporal variation of the states and the particular equations that the theory must involve? There exist historically two methods or two pictures of non-relativistic quantum mechanics. The first picture, enunciated by Schrödinger, concentrates attention on the state vectors, and the second, formulated by Heisenberg, on the dynamical variables. For many problems, these representations are equivalent; nevertheless, it is advisable to consider them separately. The equations of quantum mechanics, like any other equations of physical theory, must simply be postulated; they form an initial point of departure for problems of a new type. Let us implement these historical statements with some arguments.

Suppose that we have some vector φ. Which physical operator determines the variation of $\varphi(t)$ with time t? That is,

$$\frac{\mathrm{d}\varphi(t)}{\mathrm{d}t} = (?)\varphi(t).$$

We apply de Broglie's plane wave,

$$\varphi(t) \sim \mathrm{e}^{\mathrm{i}(\mathbf{p}\cdot\mathbf{r}-Et)/\hbar},$$

in which appear momentum $\mathbf{p}$, radius vector $\mathbf{r}$ and energy E of the particle, that must be a solution of the sought equation. We have

$$\frac{\mathrm{d}}{\mathrm{d}t}\varphi(t) = -\frac{\mathrm{i}}{\hbar}E\,\varphi(t).$$

For a freely moving particle, $E = \mathbf{p}^2/2m$, in which m is the mass of the particle, hence

$$\frac{\mathrm{d}}{\mathrm{d}t}\varphi(t) = -\frac{\mathrm{i}}{2m\hbar}\mathbf{p}^2\,\varphi(t)$$

and

$$(?)\varphi(t) = -\frac{\mathrm{i}}{2m\hbar}\mathbf{p}^2\,\varphi(t).$$

We already know the answer to this question: the squared momentum operator has a form

$$-\hbar^2\frac{\partial^2}{\partial\mathbf{r}^2};$$

$$-\hbar^2\frac{\partial^2}{\partial\mathbf{r}^2}\varphi(t) = \mathbf{p}^2\,\varphi(t),$$

and

$$(?) = \frac{1}{\mathrm{i}\hbar}\left(-\frac{\hbar^2\nabla^2}{2m}\right).$$

We see that the kinetic energy of the particle,

$$-\hbar^2\nabla^2/2m,$$

determines the sought operator. One might, thus, hope that generally

$$(?) = \frac{1}{\mathrm{i}\hbar}H;$$

i.e. the rate of variation of $\varphi(t)$ is defined by Hamiltonian H, which represents the total energy operator of the system. As a result,

$$\mathrm{i}\hbar\frac{\mathrm{d}\varphi(t)}{\mathrm{d}t} = H\,\varphi(t).$$

This equation, formulated by Schrödinger, is the principal equation of non-relativistic quantum theory. It describes the temporal variation of the states of the system that is characterized by Hamiltonian H. An additional argument of this fundamental approach is that, according to the theory of relativity, the relation between energy and time must be similar to the relation between momentum and coordinate.

For Schrödinger's equation, one generally uses the coordinate representation,

$$\mathrm{i}\hbar\frac{\partial\varphi}{\partial t} = -\frac{\hbar^2}{2}\sum_{i=1}^{N}\frac{1}{m_i}\nabla_i^2\varphi + V(t,\mathbf{r}_1,\mathbf{r}_2,\ldots,\mathbf{r}_N)\varphi,$$

$$\varphi = \varphi(t,\mathbf{r}_1,\mathbf{r}_2,\ldots,\mathbf{r}_N),$$

in which V is the operator of potential energy of interacting particles, m_i and $\mathbf{r}_i$ are the mass and radius vector of particle i, $i = 1, 2, \ldots, N$. Through the presence of Laplacians, Schrödinger's equation, in this case, is a differential equation of second order. In Cartesian coordinates (x,y,z), Laplace's operator has a form

$$\nabla^2 = \frac{\partial^2}{\partial x^2} + \frac{\partial^2}{\partial y^2} + \frac{\partial^2}{\partial z^2},$$

in spherical coordinates (r,θ,ϕ),

$$\nabla^2 = \frac{1}{r^2}\frac{\partial}{\partial r}\left(r^2\frac{\partial}{\partial r}\right) + \frac{1}{r^2}\left[\frac{1}{\sin\theta}\frac{\partial}{\partial\theta}\left(\sin\theta\frac{\partial}{\partial\theta}\right) + \frac{1}{\sin^2\theta}\frac{\partial^2}{\partial\phi^2}\right],$$

and in cylindrical coordinates (ρ,ϕ,z),

$$\nabla^2 = \frac{1}{\rho}\frac{\partial}{\partial\rho}\left(\rho\frac{\partial}{\partial\rho}\right) + \frac{1}{\rho^2}\frac{\partial^2}{\partial\phi^2} + \frac{\partial^2}{\partial z^2}.$$

In a case of stationary states,

$$\varphi(t) = \mathrm{e}^{-\mathrm{i}Et/\hbar}|\psi\rangle,$$

the energy of the system E is invariant; Schrödinger's equation becomes simplified to

$$E\mathrm{e}^{-\mathrm{i}Et/\hbar}|\psi\rangle = H(\mathrm{e}^{-\mathrm{i}Et/\hbar}|\psi\rangle);$$

and, as the Hamiltonian is independent of time,

$$H|\psi\rangle = E|\psi\rangle.$$

The latter is an equation for eigenvalues E and eigenfunctions $|\psi\rangle$ of Hamiltonian H. Eigenvalues E define possible energy levels either discrete or continuous. Using classical mechanics, one might thus select a convenient Hamiltonian; for instance, for a particle in a field with potential $V(\mathbf{r})$, it is given by the expression

$$H = \frac{\mathbf{p}^2}{2m} + V(\mathbf{r}).$$

One might furthermore replace variables by operators, e.g. in the coordinate representation, $\mathbf{r} \to \mathbf{r}$ and $\mathbf{p} \to -\mathrm{i}\hbar\nabla$; one then solves the equation for eigenvalues E and eigenfunctions $\psi(\mathbf{r})$, e.g.

$$-\frac{\hbar^2}{2m}\nabla^2\,\psi(\mathbf{r}) + V\,\psi(\mathbf{r}) = E\,\psi(\mathbf{r}),$$

and, as a result, obtains the observable energy of the system, E. Afterwards one might work with a prepared system of levels; for instance, one might determine other state vectors characterizing the transitions of the system under the influence of physical interactions from one stationary state to another. The set of stationary states is complete, such that any other state can be represented in a form of superposition of stationary states. Note that, in quantum mechanics, the coordinate and momentum are not measurable in one state, whereas energy, being a function of coordinates and momenta, might have determinate values.

Let us proceed to another picture of quantum mechanics. How did Heisenberg reason? According to Bohr's postulate, a system making a transition from state $|i\rangle$ with energy E_i to $|f\rangle$ with E_f emits a quantum with frequency ω:

$$E_i - E_f = \hbar\omega.$$

We consider a commutator of some dynamical variable A and Hamiltonian H, and calculate its matrix element,

$$\langle f|(AH - HA)|i\rangle.$$

As E_i and E_f are eigenvalues of the Hamiltonian:

$$H|i\rangle = E_i|i\rangle \text{ and } H|f\rangle = E_f|f\rangle,$$

then

$$\langle f|(AH - HA)|i\rangle = \langle f|A(H|i\rangle) - (\langle f|H)A|i\rangle = (E_i - E_f)\langle f|A|i\rangle = \hbar\omega\langle f|A|i\rangle.$$

Heisenberg supposed that the matrix element of each variable depends harmonically on time, hence

$$\langle f|A|i\rangle \sim \mathrm{e}^{-\mathrm{i}\omega t}$$

and

$$-\mathrm{i}\omega\langle f|A|i\rangle = \frac{\mathrm{d}}{\mathrm{d}t}\langle f|A|i\rangle.$$

He supposed, moreover, that the vectors are independent of time, such that

$$\left\langle f\left|\mathrm{i}\hbar\frac{\mathrm{d}A}{\mathrm{d}t}\right|i\right\rangle = \langle f|(AH - HA)|i\rangle,$$

where from

$$\mathrm{i}\hbar\frac{\mathrm{d}A}{\mathrm{d}t} = [A, H].$$

This equation, which bears Heisenberg's name, represents the equation of motion for some dynamical variable. If Heisenberg assumed that dynamical variables do not depend on time, but that the state vectors so depend, he would have arrived at Schrödinger's equation, in which specifically the state vectors depend on time, but not the dynamical variables. This distinction is principal between these two pictures of quantum mechanics.

To postulate Heisenberg's equations, it suffices, however, to apply the method of classical analogy. For an arbitrary dynamical variable A, the classical equation of motion has a form

$$\frac{\mathrm{d}A}{\mathrm{d}t} = \{A, H\};$$

in quantum mechanics, the commutator divided by $i\hbar$ corresponds to Poisson bracket $\{A,H\}$; consequently,

$$i\hbar\frac{dA}{dt} = [A, H].$$

If A depends explicitly on time, according to the classical analogy,

$$\frac{dA}{dt} = \frac{\partial A}{\partial t} + \{A, H\},$$

for a general expression of Heisenberg's equations, we obtain

$$i\hbar\frac{dA}{dt} = i\hbar\frac{\partial A}{\partial t} + [A, H].$$

Each picture is elegant in its own way. Schrödinger's equation is useful to determine the stationary states. The equations of motion in Heisenberg's form are applicable when we consider the so-called constants of motion. As a constant of motion, we understand some quantity A that satisfies the condition $dA/dt = 0$, such that A is dynamically independent of time; in this case,

$$[A, H] = 0.$$

Any constant of motion can thus be measured together with energy in one and the same state. For instance, if $A = H$ and H is explicitly independent of time, $[H, H] = 0$ and the conservation law of energy is fulfilled. If A is momentum $\mathbf{p}$ of a freely moving particle,

$$H = \frac{\mathbf{p}^2}{2m},$$

$[\mathbf{p}, H] = 0$ and the momentum is invariant.

For particular cases, Heisenberg's equations have a recognizable similarity with the equations of motion in classical mechanics. For instance, we consider the Hamiltonian of a particle in a field V

$$H = \frac{\mathbf{p}^2}{2m} + V(\mathbf{r}),$$

and calculate $d\mathbf{p}/dt$. We have

$$i\hbar\frac{d\mathbf{p}}{dt} = [\mathbf{p}, H] = [\mathbf{p}, V(\mathbf{r})] = -i\hbar\frac{\partial V}{\partial \mathbf{r}},$$

where from

$$\frac{d\mathbf{p}}{dt} = -\frac{\partial V}{\partial \mathbf{r}}.$$

This is Newton's equation for a motion of the particle in a potential field, but already in operator form.

Another example is a calculation of velocity $\mathbf{v} = \mathrm{d}\mathbf{r}/\mathrm{d}t$. In an analogous manner, we have

$$\mathrm{i}\hbar\frac{\mathrm{d}\mathbf{r}}{\mathrm{d}t} = [\mathbf{r}, H] = \frac{1}{2m}\left[\mathbf{r}, \mathbf{p}^2\right] = \mathrm{i}\hbar\frac{\mathbf{p}}{m},$$

where from

$$\frac{\mathrm{d}\mathbf{r}}{\mathrm{d}t} = \frac{\mathbf{p}}{m}.$$

Classical and quantum-mechanical definitions of the velocity thus coincide.

These coincidences, being purely formal, certainly confirm the correctness of Heisenberg's conclusions. In quantum mechanics, it is not the dynamical variables that have physical meaning but their eigenvalues, which are determined from other equations. If we set $\hbar$ equal to zero and assume that, in this case, all dynamical variables become commutative quantities with each other, the equations for dynamical variables (linear operators) and observables (eigenvalues) become coincident in an absolute manner.

Dirac's Theory

Despite all the successes of Schrödinger's non-relativistic theory, it is physically unsatisfactory: it fails to explain the spin of the electron, to yield the correct expression for fine structure and to take into account the specification of quantum-electrodynamic effects. According to Dirac, the principal problem of the *old* theory involves how to choose a Hamiltonian H. In a non-relativistic case,

$$H = \frac{\mathbf{p}^2}{2m} + \cdots,$$

in which m is the mass of a particle and $\mathbf{p}$ is its momentum; in Schrödinger's equation, there is therefore no symmetry between space coordinates and time t, between quantity of energy

$$E \rightarrow \mathrm{i}\hbar\frac{\partial}{\partial t}$$

and components of momentum p_x, p_y and p_z. To increase the attraction to quantum theory, one should either combine Schrödinger's equation with a relativistic Hamiltonian or discover another Hamiltonian altogether.

We consider the former scheme. Momentum $\mathbf{p}$ and energy E of a particle are related to each other, forming a four-vector

$$p^{\mu} = \left(\frac{E}{c}, p_x, p_y, p_z\right), \quad \mu = 0, 1, 2, 3,$$

such that

$$\sum_{\mu,\nu} g^{\mu\nu} p_\mu p_\nu = \sum_{\mu,\nu} g_{\mu\nu} p^\mu p^\nu \equiv p_\mu p^\mu = \left(\frac{E}{c}\right)^2 - \mathbf{p}^2 = m^2c^2,$$

in which c is the speed of light and

$$g^{\mu\nu} = \begin{pmatrix} 1 & 0 & 0 & 0 \\ 0 & -1 & 0 & 0 \\ 0 & 0 & -1 & 0 \\ 0 & 0 & 0 & -1 \end{pmatrix}$$

is Minkowski's metric tensor. In the classical expression,

$$(E/c)^2 - \mathbf{p}^2 = m^2c^2,$$

replacing E, according to Schrödinger's equation, with operator $i\hbar\, \partial/\partial t$, and $\mathbf{p}$ with operator $-i\hbar\nabla$, we obtain the equation,

$$\left(\hbar^2 \frac{\partial^2}{\partial t^2} - \hbar^2 c^2 \nabla^2 + m^2c^4\right)\psi = 0$$

or

$$(p_\mu p^\mu - m^2c^2)\psi = 0,$$

in which ψ is the wave function of the particle, $p^\mu = i\hbar\, \partial/\partial x_\mu$ and $x_\mu = (ct, -\mathbf{r})$;

$$x^\mu = \sum_\nu g^{\mu\nu} x_\nu = (ct, \mathbf{r}) \text{ and } p_\mu = \sum_\nu g_{\mu\nu} p^\nu = (p_0, -\mathbf{p}).$$

Quantity

$$p_0 = p^0 = i\hbar \frac{\partial}{\partial x_0} = \frac{i\hbar}{c}\frac{\partial}{\partial t} \to \frac{E}{c}$$

represents a fourth temporal component of the momentum operator.

As we see, this first scheme to construct the relativistic quantum theory fails to become sufficiently informative; it yields a solution with a negative value of energy,

$$E = \pm\sqrt{\mathbf{p}^2c^2 + m^2c^4}.$$

Whether this situation is unsatisfactory or not becomes clear when, together with particles, antiparticles come under consideration. The obtained equation, which bears the names of Klein, Fock and Gordon, is relativistically invariant and applicable to describe a particle with spin that equals zero. Schrödinger also obtained this equation.

Following Dirac, we consider the second scheme to modify Schrödinger's equation, which amounts to a search for a Hamiltonian of a new type. Substituting $i\hbar\, \partial/\partial t$ by cp_0, we have

$$p_0\psi = \frac{H}{c}\psi \equiv (?)\psi.$$

As p_0 enters into the equation linearly, one expects other components of the four-vector of momentum to appear in the equation in a linear manner. Hence,

$$(?)\psi = (\alpha_1 p_1 + \alpha_2 p_2 + \alpha_3 p_3 + \beta)\psi,$$

such that

$$(p_0 - \alpha_1 p_1 - \alpha_2 p_2 - \alpha_3 p_3 - \beta)\psi = 0,$$

in which quantities α and β are independent of neither coordinates nor momenta; they describe the new degrees of freedom that are hidden from classical mechanics. We multiply this equation by $(p_0 + \alpha_1 p_1 + \alpha_2 p_2 + \alpha_3 p_3 + \beta)$ on the left,

$$\left(p_0^2 - \sum_r \alpha_r^2 p_r^2 - \beta^2 - \sum_{r \neq s}(\alpha_r\alpha_s + \alpha_s\alpha_r)p_r p_s - \sum_r (\alpha_r\beta + \beta\alpha_r)p_r\right)\psi = 0;$$

to bring the latter into coincidence with equation $(p_0^2 - \mathbf{p}^2 - m^2c^2)\psi = 0$, one must put

$$\begin{aligned} \alpha_r^2 = 1, \quad \beta^2 = m^2c^2, \\ \alpha_r\alpha_s + \alpha_s\alpha_r = 0 \quad \text{at } r \neq s, \\ \alpha_r\beta + \beta\alpha_r = 0. \end{aligned}$$

These relations are the equivalent of the well-known rules for Pauli matrices,

$$\sigma_r\sigma_s + \sigma_s\sigma_r = 2\delta_{rs},$$

in which

$$\sigma_1 = \begin{pmatrix} 0 & 1 \\ 1 & 0 \end{pmatrix}, \quad \sigma_2 = \begin{pmatrix} 0 & -\mathrm{i} \\ \mathrm{i} & 0 \end{pmatrix}, \quad \sigma_3 = \begin{pmatrix} 1 & 0 \\ 0 & -1 \end{pmatrix}.$$

We must have, however, four matrices, not three; matrices 2×2 are therefore insufficient for our purpose. Let us determine the minimum size N of the new matrices. As, for instance, $\alpha_1\alpha_2 = -\alpha_2\alpha_1$,

$$\det \alpha_1 \cdot \det \alpha_2 = \det(-I) \cdot \det \alpha_2 \cdot \det \alpha_1,$$

where from

$$\det(-I) = (-1)^N = 1.$$

Number N is thus even and equal to at least four. Moreover, $\alpha_2 = -\alpha_1^{-1}\alpha_2\alpha_1$, such that

$$\mathrm{Sp}(\alpha_2) = -\mathrm{Sp}(\alpha_1^{-1}\alpha_2\alpha_1) = -\mathrm{Sp}(\alpha_2) = 0;$$

spurs $\mathrm{Sp}(\alpha_1)$, $\mathrm{Sp}(\alpha_3)$ and $\mathrm{Sp}(\beta)$ also equal zero.

To satisfy all these relations with regard to $\boldsymbol{\alpha}$ and $\boldsymbol{\beta}$, we extend the system of Pauli matrices in a diagonal manner

$$\boldsymbol{\sigma} \to \begin{pmatrix} \boldsymbol{\sigma} & 0 \\ 0 & \boldsymbol{\sigma} \end{pmatrix},$$

i.e.

$$\sigma_1 = \begin{pmatrix} 0 & 1 & 0 & 0 \\ 1 & 0 & 0 & 0 \\ 0 & 0 & 0 & 1 \\ 0 & 0 & 1 & 0 \end{pmatrix}, \quad \sigma_2 = \begin{pmatrix} 0 & -\mathrm{i} & 0 & 0 \\ \mathrm{i} & 0 & 0 & 0 \\ 0 & 0 & 0 & -\mathrm{i} \\ 0 & 0 & \mathrm{i} & 0 \end{pmatrix}, \quad \sigma_3 = \begin{pmatrix} 1 & 0 & 0 & 0 \\ 0 & -1 & 0 & 0 \\ 0 & 0 & 1 & 0 \\ 0 & 0 & 0 & -1 \end{pmatrix}.$$

We introduce three more matrices ρ_1, ρ_2 and ρ_3, having interchanged in σ_r the second and third rows and columns:

$$\boldsymbol{\rho} = \left\{ \begin{pmatrix} 0 & I \\ I & 0 \end{pmatrix}, \ \mathrm{i}\begin{pmatrix} 0 & -I \\ I & 0 \end{pmatrix}, \begin{pmatrix} I & 0 \\ 0 & -I \end{pmatrix} \right\},$$

in which I is a 2×2 unit matrix. As we see, $\boldsymbol{\rho}$ has the structure of Pauli matrices with 2×2 elements; obviously

$$\rho_r\rho_s + \rho_s\rho_r = 2\delta_{rs} \ \text{ and } \ \rho_r\sigma_s = \sigma_s\rho_r.$$

According to Dirac, we assume

$$\alpha_r = \rho_1\sigma_r \ \text{ and } \ \beta = mc\rho_3;$$

accordingly, with this definition,

$$\alpha_r^2 = \rho_1^2\sigma_r^2 = 1,$$
$$\alpha_1\alpha_2 = \rho_1^2\sigma_1\sigma_2 = -\rho_1^2\sigma_2\sigma_1 = -\alpha_2\alpha_1$$

and so on.

As a result,

$$(p_0 - \rho_1(\boldsymbol{\sigma} \cdot \mathbf{p}) - mc\rho_3)\psi = 0.$$

This equation, first derived by Dirac, describes particles with spin equal to one half. To rewrite the new equation in a covariant manner, we multiply it by ρ_3,

$$(\rho_3 p_0 - \rho_3\rho_1(\boldsymbol{\sigma} \cdot \mathbf{p}) - mc)\psi = 0,$$

put, by definition, $\gamma^0 = \rho_3$ and $\gamma^r = \rho_3\rho_1\sigma_r$; consequently,

$$(\gamma^\mu p_\mu - mc)\psi = 0;$$

Latin indices correspond to three-vector, and Greek indices to four-vector. Dirac's matrices have an explicit form

$$\gamma^0 = \begin{pmatrix} I & 0 \\ 0 & -I \end{pmatrix} \text{ and } \boldsymbol{\gamma} = \begin{pmatrix} 0 & \boldsymbol{\sigma} \\ -\boldsymbol{\sigma} & 0 \end{pmatrix}.$$

In modern quantum theory, together with Schrödinger's and Heisenberg's pictures, Dirac's equation has a place similar to those of Lagrange's equations in mechanics and Maxwell's equations in electrodynamics. In the new wave equation, the relativistic structure and the rules of non-commutative algebra are naturally combined; there is no problem concerned with the negativity of the density of states, and the principal results are experimentally confirmed. The conformation to the theory of relativity demands, however, additional elucidation. The new theory must yield results that are independent of the choice of a Lorentz frame of reference.

We consider a linear transformation from x^ν to x'^μ:

$$x^\nu = a^{\nu\mu}x'_\mu, \quad x'^\mu = a^{\mu\nu}x_\nu, \quad a_{\mu\nu}a^{\nu\tau} = \delta^\tau_\mu.$$

Suppose that Dirac's equation written in the new coordinates retains its initial form, i.e.

$$\left(i\hbar\gamma^\mu \frac{\partial}{\partial x'^\mu} - mc\right)\psi' = 0,$$

in which ψ' is a function of coordinates x'^μ. With the aid of this transformation,

$$\psi' = S\psi,$$

we return to the initial variables. We have

$$\frac{\partial}{\partial x'^\mu} = \frac{\partial x^\nu}{\partial x'^\mu} \cdot \frac{\partial}{\partial x^\nu} = a^\nu_\mu \cdot \frac{\partial}{\partial x^\nu},$$

such that

$$\left(i\hbar a_{\mu}^{\nu}\gamma^{\mu}S\frac{\partial}{\partial x^{\nu}}-Smc\right)\psi=0.$$

Through an orthogonality of transformation,

$$S^{-1}S=1,$$

multiplying the obtained equation by S^{-1} on the left side, we consequently find

$$\left(i\hbar(S^{-1}a_{\mu}^{\nu}\gamma^{\mu}S)\frac{\partial}{\partial x^{\nu}}-mc\right)\psi=0.$$

For this equation to coincide with Dirac's equation written with primed coordinates, one must enforce the equality

$$S^{-1}a_{\mu}^{\nu}\gamma^{\mu}S=\gamma^{\nu},$$

or

$$S\gamma^{\nu}S^{-1}=a_{\mu}^{\nu}\gamma^{\mu}.$$

To prove the Lorentz invariance, we must answer two questions. Does there exist a transformation S that preserves the form of the initial Dirac equation? Might matrix S imply a Lorentz transformation matrix?

We initially reply to the first query. We consider a linear rotational transformation, for instance, in plane x_1x_2. An expression for the rotation matrix is given in a form

$$a_{\mu}^{\nu}=\begin{pmatrix}1&0&0&0\\0&\cos\phi&\sin\phi&0\\0&-\sin\phi&\cos\phi&0\\0&0&0&1\end{pmatrix},\quad\begin{cases}x_1'=x_1\cos\phi+x_2\sin\phi,\\x_2'=-x_1\sin\phi+x_2\cos\phi.\end{cases}$$

To show that $S=\exp(\phi\gamma^1\gamma^2/2)$, we have

$$S=1+\frac{\phi}{2}\gamma^1\gamma^2+\frac{\phi^2}{2!\cdot4}(\gamma^1\gamma^2)^2+\frac{\phi^3}{3!\cdot8}(\gamma^1\gamma^2)^3+\frac{\phi^4}{4!\cdot16}(\gamma^1\gamma^2)^4+\cdots;$$

as

$$\begin{aligned}(\gamma^1\gamma^2)^2&=\gamma^1\gamma^2\gamma^1\gamma^2=-(\gamma^1)^2(\gamma^2)^2=-1,\\(\gamma^1\gamma^2)^3&=(\gamma^1\gamma^2)^2\gamma^1\gamma^2=-\gamma^1\gamma^2,\\(\gamma^1\gamma^2)^4&=+1\end{aligned}$$

and so on,

$$S = \left(1 - \frac{\phi^2}{2! \cdot 4} + \frac{\phi^4}{4! \cdot 16} - \cdots\right) + \gamma^1\gamma^2\left(\frac{\phi}{2} - \frac{\phi^3}{3! \cdot 8} + \cdots\right) = \cos\frac{\phi}{2} + \gamma^1\gamma^2 \sin\frac{\phi}{2}.$$

One readily verifies that $S^{-1}S = 1$ if

$$S^{-1} = \exp\left(\frac{-\phi\gamma^1\gamma^2}{2}\right) = \cos\frac{\phi}{2} - \gamma^1\gamma^2 \sin\frac{\phi}{2}.$$

Finally,

$$S\gamma^\nu S^{-1} = \gamma^\nu \cos^2\left(\frac{\phi}{2}\right) - \gamma^\nu\gamma^1\gamma^2 \cos\left(\frac{\phi}{2}\right) \sin\left(\frac{\phi}{2}\right) + \gamma^1\gamma^2\gamma^\nu \cos\left(\frac{\phi}{2}\right) \sin\left(\frac{\phi}{2}\right) - \gamma^1\gamma^2\gamma^\nu\gamma^1\gamma^2 \sin^2\left(\frac{\phi}{2}\right),$$

where from

$$\begin{aligned} S\gamma^1 S^{-1} &= \gamma^1 \cos\phi + \gamma^2 \sin\phi, \\ S\gamma^2 S^{-1} &= -\gamma^1 \sin\phi + \gamma^2 \cos\phi, \\ S\gamma^\nu S^{-1} &= \gamma^\nu \text{ at } \nu \neq 1, 2. \end{aligned}$$

One sees that $S\gamma^\nu S^{-1} = a^\nu_\mu \gamma^\mu$, so that Dirac's equation is indeed invariant with regard to a rotational transformation.

There is then no major difficulty to answer the second question and to prove Lorentz invariance. A Lorentz transformation – a conversation to the system moving with regard to an initial system with velocity $v = \text{constant}$ – is well known to represent a rotation in plane x_1x_0 by an imaginary angle. Putting $\phi = i\vartheta$ and bearing in mind the imaginary unit at the temporal coordinate x_0, we have

$$a^\nu_\mu = \begin{pmatrix} \text{ch}\,\vartheta & -\text{sh}\,\vartheta & 0 & 0 \\ -\text{sh}\,\vartheta & \text{ch}\,\vartheta & 0 & 0 \\ 0 & 0 & 1 & 0 \\ 0 & 0 & 0 & 1 \end{pmatrix}, \quad \begin{cases} x'_0 = x_0\,\text{ch}\,\vartheta - x_1\,\text{sh}\,\vartheta, & \text{th}\,\vartheta = v/c, \\ x'_1 = -x_0\,\text{sh}\,\vartheta + x_1\,\text{ch}\,\vartheta, & \text{ch}\,\vartheta = (1-(v/c)^2)^{-1/2}. \end{cases}$$

The sought transformation matrix acquires a form

$$S = \exp\left(\frac{i\vartheta\gamma^0\gamma^1}{2}\right) = \text{ch}\left(\frac{\vartheta}{2}\right) + i\gamma^0\gamma^1\text{sh}\left(\frac{\vartheta}{2}\right);$$

obviously,

$$S^{-1} = \exp\left(\frac{-i\vartheta\gamma^0\gamma^1}{2}\right) = \text{ch}\left(\frac{\vartheta}{2}\right) - i\gamma^0\gamma^1\text{sh}\left(\frac{\vartheta}{2}\right).$$

The wave equation of Dirac is thus relativistically invariant; it yields physical results that are independent of the Lorentz frame of reference.

Spin and Magnetic Moment

The wave equation of Dirac is essential to explain the doubling of stationary levels for an electron in an atom. According to Schrödinger's picture, one might circumvent this difficulty through a phenomenological introduction of an electron spin that equals $\hbar/2$ and a magnetic moment equal to the Bohr magneton $\mu_B = e\hbar/2mc$. Although Pauli, having heuristically applied this approach, succeeded in treating the new phenomenon, the nature of the pertinent degrees of freedom remained unclear. In this sense, Dirac's theory elucidated all aspects in question. Apart from an experimental confirmation, the spin and magnetic moment of the electron have acquired a solid theoretical foundation.

We extrapolate Dirac's equation to the case of the presence of an external electromagnetic field. As in classical physics, one should replace energy E with $E + eA_0$ and momentum $\mathbf{p}$ with $\mathbf{p} + e\mathbf{A}/c$, in which e is the absolute value of an electronic charge, c is the speed of light and A_0 and $\mathbf{A}$ are corresponding scalar and vector potentials of a field. If $A^\mu = (A^0, \mathbf{A})$ is the four-vector of a field potential,

$$p_\mu \rightarrow p_\mu + \frac{e}{c} A_\mu$$

This replacement possesses both gradient and Lorentz invariance. As a result, we obtain

$$\left(\gamma^\mu \left(p_\mu + \frac{e}{c} A_\mu\right) - mc\right)\psi = 0,$$

in which m is the mass of the electron. We multiply the obtained equation by $\gamma^\nu(p_\nu + eA_\nu/c)$ on the left to yield

$$\left(\gamma^\nu \gamma^\mu \left(p_\nu + \frac{e}{c} A_\nu\right)\left(p_\mu + \frac{e}{c} A_\mu\right) - m^2c^2\right)\psi = 0.$$

One sees that γ-matrices of Dirac satisfy the relation of Clifford algebra

$$\gamma_\mu \gamma_\nu + \gamma_\nu \gamma_\mu = 2g_{\mu\nu}.$$

If this relation is combined with an antisymmetric tensor

$$\sigma^{\nu\mu} = -\sigma^{\mu\nu} = \frac{\mathrm{i}}{2}(\gamma^\nu \gamma^\mu - \gamma^\mu \gamma^\nu),$$

one might directly express the product $\gamma^\nu \gamma^\mu$ through $g^{\nu\mu}$ and $\sigma^{\nu\mu}$:

$$\gamma^\nu \gamma^\mu = g^{\nu\mu} - \mathrm{i}\sigma^{\nu\mu}.$$

Consequently,

$$\gamma^\nu\gamma^\mu\left(p_\nu + \frac{eA_\nu}{c}\right)\left(p_\mu + \frac{eA_\mu}{c}\right)$$
$$= \left(p^\mu + \frac{eA^\mu}{c}\right)\left(p_\mu + \frac{eA_\mu}{c}\right) - \mathrm{i}\sigma^{\nu\mu}\left(p_\nu + \frac{eA_\nu}{c}\right)\left(p_\mu + \frac{eA_\mu}{c}\right)$$
$$= \left(p^\mu + \frac{eA^\mu}{c}\right)\left(p_\mu + \frac{eA_\mu}{c}\right) - \frac{\mathrm{i}}{2}\left(\sigma^{\nu\mu} - \sigma^{\mu\nu}\right)\left(p_\nu + \frac{eA_\nu}{c}\right)\left(p_\mu + \frac{eA_\mu}{c}\right)$$
$$= \left(p^\mu + \frac{eA^\mu}{c}\right)\left(p_\mu + \frac{eA_\mu}{c}\right) - \frac{\mathrm{i}}{2}\sigma^{\nu\mu}\left[p_\nu + \frac{eA_\nu}{c}, p_\mu + \frac{eA_\mu}{c}\right].$$

Here, for the commutator, we have

$$\left[p_\nu + \frac{eA_\nu}{c}, p_\mu + \frac{eA_\mu}{c}\right] = \left[p_\nu, \frac{eA_\mu}{c}\right] - \left[p_\mu, \frac{eA_\nu}{c}\right] = \frac{\mathrm{i}e\hbar}{c}\left(\frac{\partial A_\mu}{\partial x^\nu} - \frac{\partial A_\nu}{\partial x^\mu}\right) = \frac{\mathrm{i}e\hbar}{c}F_{\nu\mu},$$

in which

$$F_{\nu\mu} = \frac{\partial A_\mu}{\partial x^\nu} - \frac{\partial A_\nu}{\partial x^\mu} = \begin{pmatrix} 0 & E_1 & E_2 & E_3 \\ -E_1 & 0 & -B_3 & B_2 \\ -E_2 & B_3 & 0 & -B_1 \\ -E_3 & -B_2 & B_1 & 0 \end{pmatrix}$$

is the tensor of the electromagnetic field with a polar electric-field vector $\mathbf{E} = (E_1, E_2, E_3)$ and an axial magnetic-field vector $\mathbf{B} = (B_1, B_2, B_3)$.

The quadratic Dirac's equation in the external field acquires a form

$$\left(\left(p^\mu + \frac{e}{c}A^\mu\right)\left(p_\mu + \frac{e}{c}A_\mu\right) + \frac{e\hbar}{2c}\sigma^{\nu\mu}F_{\nu\mu} - m^2c^2\right)\psi = 0.$$

To simplify it, we put $\gamma_r = \rho_3\rho_1\sigma_r = \mathrm{i}\rho_2\sigma_r$ and $\gamma_0 = \rho_3$. We have

$$\sigma^{0r} = \frac{1}{2}\sigma^r(\rho_2\rho_3 - \rho_3\rho_2) = \mathrm{i}\rho_1\sigma^r = \mathrm{i}\alpha^r,$$

$$\sigma^{12} = -\frac{\mathrm{i}}{2}(\sigma^1\sigma^2 - \sigma^2\sigma^1) = -\mathrm{i}\sigma^1\sigma^2 = \sigma^3, \quad \sigma^{23} = \sigma^1, \quad \sigma^{31} = \sigma^2.$$

Consequently,

$$\sigma^{\nu\mu}F_{\nu\mu} = 2\sigma^{0r}E_r + 2\sigma^{rs}F_{rs}|_{r<s} = 2\mathrm{i}\alpha^r E_r - 2\sigma^r B_r,$$

and the equation for an electron in an external field becomes

$$\left(\left(\frac{E+eA_0}{c}\right)^2 - \left(\mathbf{p}+\frac{e\mathbf{A}}{c}\right)^2 - \frac{e\hbar}{c}(\boldsymbol{\sigma}\cdot\mathbf{B}) + \mathrm{i}\frac{e\hbar}{c}(\boldsymbol{\alpha}\cdot\mathbf{E}) - m^2c^2\right)\psi = 0.$$

Here, we perceive two supplementary terms

$$-\frac{e\hbar}{c}(\boldsymbol{\sigma}\cdot\mathbf{B}) \text{ and } \mathrm{i}\frac{e\hbar}{c}(\boldsymbol{\alpha}\cdot\mathbf{E}).$$

The former shows the presence of the new degree of freedom for an electron – spin and the magnetic moment concerned with spin,

$$\boldsymbol{\mu} = -\frac{e\hbar\boldsymbol{\sigma}}{2mc}$$

that interacts with an external magnetic field $\mathbf{B}$. Spin emphasizes that an electron, possessing an inner mechanical angular momentum, 'rotates' about its own axis. The latter term turns out to be imaginary; its principal purpose is to ensure the relativistic invariance of Dirac's theory.

According to a physical point of view, the purely imaginary term is of only minor interest because it corresponds to the presence of an imaginary electric moment for the electron. One might suppose that its appearance is necessary only for that purpose, in a formal manner, to adapt the new theory to Schrödinger's picture. The latter is essentially non-relativistic, and its role is therefore highly doubtful. Omitting this imaginary term, we define this non-relativistic limit. One should assume that

$$E = \varepsilon + mc^2, \quad eA_0 \ll mc^2 \text{ and } \varepsilon \ll mc^2,$$

then

$$\left(\frac{E+eA_0}{c}\right)^2 - m^2c^2 \approx 2m(\varepsilon + eA_0),$$

and

$$\left(\frac{1}{2m}\left(\mathbf{p}+\frac{e\mathbf{A}}{c}\right)^2 - eA_0 + \frac{e\hbar}{2mc}(\boldsymbol{\sigma}\cdot\mathbf{B})\right)\psi = \varepsilon\psi,$$

which constitutes the famous Pauli equation.

Applying another consideration, one might arrive at a definition of spin. The angular momentum in a central field of force, for which $\mathbf{A} = 0$ and $A_0 = A_0(r)$, is invariant. In this case, Dirac's Hamiltonian, additionally multiplied by c, has a form

$$H = -eA_0(r) + c\rho_1(\boldsymbol{\sigma}\cdot\mathbf{p}) + \rho_3 mc^2.$$

We calculate commutator [**L**,H], in which $\mathbf{L} = (L_1,L_2,L_3)$ is the orbital angular momentum of the electron. We have

$$\begin{aligned}[L_1, H] &= c\rho_1 \boldsymbol{\sigma} \cdot [L_1, \mathbf{p}] = c\rho_1 \boldsymbol{\sigma} \cdot (\mathbf{j}[L_1, p_2] + \mathbf{k}[L_1, p_3]) \\ &= \mathrm{i}\hbar c\rho_1(\sigma_2 p_3 - \sigma_3 p_2) = \mathrm{i}\hbar c\rho_1(\boldsymbol{\sigma} \times \mathbf{p})_1;\end{aligned}$$

consequently, $[\mathbf{L},H] = \mathrm{i}\hbar c\rho_1(\boldsymbol{\sigma} \times \mathbf{p})$ and angular momentum **L** fails to be invariant. We proceed to calculate [$\boldsymbol{\sigma}$,H]:

$$\begin{aligned}[\sigma_1, H] &= c\rho_1[\sigma_1, \boldsymbol{\sigma}] \cdot \mathbf{p} = c\rho_1(\mathbf{j}[\sigma_1, \sigma_2] + \mathbf{k}[\sigma_1, \sigma_3]) \cdot \mathbf{p} \\ &= 2\mathrm{i}c\rho_1(\sigma_3 p_2 - \sigma_2 p_3) = -2\mathrm{i}c\rho_1(\boldsymbol{\sigma} \times \mathbf{p})_1;\end{aligned}$$

thus, $[\hbar\boldsymbol{\sigma}/2,H] = -\mathrm{i}\hbar c\rho_1(\boldsymbol{\sigma} \times \mathbf{p})$. One sees that

$$\left[\mathbf{L} + \frac{\hbar}{2}\boldsymbol{\sigma}, H\right] = 0,$$

such that vector $\mathbf{L} + \hbar\boldsymbol{\sigma}/2$ is a constant of the motion. The electron thus possesses an inner angular momentum $\hbar\boldsymbol{\sigma}/2$, which is appropriately called spin. The eigenvalues of one projection of quantity $\boldsymbol{\sigma}$ equal ± 1, which conforms entirely to the hypothesis of Goudsmit and Uhlenbeck; the observable values of spin momentum are $\pm\hbar/2$. Spin is an exceptional quantum quantity that tends to zero in a classical limit as $\hbar \to 0$.

The Pauli equation derived above is the result of a particular non-relativistic limit for Dirac's theory. Dirac's equation admits, however, another cardinal non-relativistic consideration that yields physically correct results with no additional supposition, unlike what Pauli's phenomenological theory includes. To investigate this limiting case, we write Dirac's equation in an external electric field with potential A_0:

$$\left(\left(p_0 + \frac{eA_0}{c}\right) - \rho_1(\boldsymbol{\sigma} \cdot \mathbf{p}) - \rho_3 mc\right)\psi = 0,$$

or in an explicit form after multiplying by c:

$$\left((E + eA_0)\begin{pmatrix} I & 0 \\ 0 & I \end{pmatrix} - c(\boldsymbol{\sigma} \cdot \mathbf{p})\begin{pmatrix} 0 & I \\ I & 0 \end{pmatrix} - mc^2\begin{pmatrix} I & 0 \\ 0 & -I \end{pmatrix}\right)\begin{pmatrix} \psi_A \\ \psi_B \end{pmatrix} = 0.$$

Here,

$$\psi_A = \begin{pmatrix} \psi_a \\ \psi_{a'} \end{pmatrix} \text{ and } \psi_B = \begin{pmatrix} \psi_b \\ \psi_{b'} \end{pmatrix}$$

are two-component wave functions. This equation is equivalent to a system

$$c(\boldsymbol{\sigma}\cdot\mathbf{p})\psi_B + mc^2\psi_A = (E + eA_0)\psi_A,$$
$$c(\boldsymbol{\sigma}\cdot\mathbf{p})\psi_A - mc^2\psi_B = (E + eA_0)\psi_B.$$

We assume $\varepsilon = E - mc^2$, isolate ψ_B from the latter equation and substitute it into the former to yield

$$\psi_B = c(\varepsilon + eA_0 + 2mc^2)^{-1}(\boldsymbol{\sigma}\cdot\mathbf{p})\psi_A$$

and

$$\left(\frac{1}{2m}(\boldsymbol{\sigma}\cdot\mathbf{p})\left(1+\frac{\varepsilon+eA_0}{2mc^2}\right)^{-1}(\boldsymbol{\sigma}\cdot\mathbf{p}) - eA_0\right)\psi_A = \varepsilon\psi_A.$$

In the non-relativistic case,

$$\mathbf{p} = m\mathbf{v},\quad \varepsilon \ll mc^2 \text{ and } eA_0 \ll mc^2,$$

such that

$$\psi_B \sim \frac{v}{c}|\boldsymbol{\sigma}|\psi_A$$

and two components

$$\psi_B = \begin{pmatrix}\psi_b\\ \psi_{b'}\end{pmatrix}$$

are appropriately called small. To define the large components ψ_A, we use the approximation

$$\left(1+\frac{\varepsilon+eA_0}{2mc^2}\right)^{-1} \approx 1 - \frac{\varepsilon+eA_0}{2mc^2},$$

take into account that

$$\mathbf{p}A_0 = A_0\mathbf{p} - i\hbar\frac{\partial A_0}{\partial \mathbf{r}},$$

and note equalities

$$(\boldsymbol{\sigma}\cdot\mathbf{p})^2 = \mathbf{p}^2$$

and

$$\left(\boldsymbol{\sigma}\cdot\frac{\partial A_0}{\partial \mathbf{r}}\right)(\boldsymbol{\sigma}\cdot\mathbf{p}) = \frac{\partial A_0}{\partial \mathbf{r}}\cdot\mathbf{p} + \mathrm{i}\boldsymbol{\sigma}\left(\frac{\partial A_0}{\partial \mathbf{r}}\times\mathbf{p}\right),$$

which follow from the well-known relation

$$(\boldsymbol{\sigma}\cdot\mathbf{a})(\boldsymbol{\sigma}\cdot\mathbf{b}) = \mathbf{a}\cdot\mathbf{b} + \mathrm{i}\boldsymbol{\sigma}\cdot(\mathbf{a}\times\mathbf{b}),$$

which is satisfied for arbitrary vectors **a** and **b** as a pair. Consequently,

$$(\boldsymbol{\sigma}\cdot\mathbf{p})A_0(\boldsymbol{\sigma}\cdot\mathbf{p}) = A_0\mathbf{p}^2 - \mathrm{i}\hbar\left(\frac{\partial A_0}{\partial \mathbf{r}}\mathbf{p} + \mathrm{i}\boldsymbol{\sigma}\left(\frac{\partial A_0}{\partial \mathbf{r}}\times\mathbf{p}\right)\right).$$

Supposing spherical symmetry for potential A_0, we have

$$\frac{\partial A_0}{\partial \mathbf{r}} = A_0'\frac{\mathbf{r}}{r}.$$

Thus,

$$\left(\frac{\mathbf{p}^2}{2m} - \left(\frac{\varepsilon + eA_0}{2mc^2}\right)\frac{\mathbf{p}^2}{2m} - eA_0 + \frac{\mathrm{i}e\hbar}{4m^2c^2}\left(A_0'\frac{\mathbf{r}\cdot\mathbf{p}}{r} + \mathrm{i}A_0'\boldsymbol{\sigma}\cdot\frac{\mathbf{r}\times\mathbf{p}}{r}\right)\right)\psi_A = \varepsilon\psi_A.$$

Noting that

$$\frac{\varepsilon + eA_0}{2mc^2} \approx \frac{1}{2mc^2}\cdot\frac{\mathbf{p}^2}{2m},$$

we eventually obtain

$$\left(\frac{\mathbf{p}^2}{2m} - \frac{\mathbf{p}^4}{8m^3c^2} - eA_0 + \frac{\mathrm{i}e\hbar A_0'}{4m^2c^2}\frac{\mathbf{r}\cdot\mathbf{p}}{r} - \frac{eA_0'}{2m^2c^2r}\mathbf{s}\cdot\mathbf{L}\right)\psi_A = \varepsilon\psi_A,$$

in which $\mathbf{s} = \hbar\boldsymbol{\sigma}/2$ is spin and $\mathbf{L} = \mathbf{r}\times\mathbf{p}$ is the orbital angular momentum of the electron. This scenario to proceed to the non-relativistic limit was outlined by Dirac.

According to an interpretation of the obtained equation, the first two terms follow from a classical expansion

$$\varepsilon = \sqrt{m^2c^4 + c^2\mathbf{p}^2} - mc^2 = \frac{\mathbf{p}^2}{2m} - \frac{\mathbf{p}^4}{8m^3c^2} + \cdots$$

that represents the kinetic energy of the electron. The third term $-eA_0$ is the potential energy of interaction with an external electric field. Quantity

$$\frac{ie\hbar}{4m^2c^2}A_0'\frac{\mathbf{r}\cdot\mathbf{p}}{r}=\frac{e\hbar^2}{4m^2c^2}A_0'\frac{\partial}{\partial r}$$

has no classical analogue. The latter term

$$-\frac{eA_0'}{2m^2c^2r}\mathbf{s}\cdot\mathbf{L}$$

describes a spin–orbital interaction important for physics; factor 1/2 appears here in a natural manner, not artificially as the theory of Pauli and Darwin yields. According to a phenomenological consideration, in the non-relativistic theory one might also introduce this spin–orbital coupling; for agreement with experiment, one must include by hand the so-called Thomas factor 1/2. After taking this factor into account, the theory of Pauli and Darwin allows one to obtain the correct equation, which is in agreement with experiment.

Phenomenological Description

In the experiment of Stern and Gerlach, atoms of silver in a narrow beam passed through a region of strong and inhomogeneous magnetic field. Each atom acquired additional energy $W=-\boldsymbol{\mu}\cdot\mathbf{B}$, in which $\boldsymbol{\mu}$ is the magnetic moment of the atom and $\mathbf{B}$ is the magnetic-field vector. As a result of the experiment, on a screen, Stern and Gerlach might have obtained some diffuse image corresponding to a mutual orientation $\boldsymbol{\mu}$ and $\mathbf{B}$. This result was not observed, however; instead, the atomic beam became split such that, on the screen, there were discovered only two images symmetrically disposed with respect to the initial beam. Atomic rays of alkali metals also had two images; for beams containing atoms of vanadium or manganese or iron, the number of images became more than two.

A beam of hydrogen atoms, which are in an S-state, attracts special interest. In this case, the orbital quantum number ℓ of the electron equals zero; consequently, for the electron, the mechanical angular momentum and the magnetic moment, associated with this angular momentum, are completely lacking. As a result of an experiment, the atomic beam again became split into two components under the influence of the magnetic field; this fact bears witness to two possible orientations for the magnetic moment of the electron. Uhlenbeck and Goudsmit supposed a posteriori that the electron possesses an intrinsic angular momentum – spin, and the projection of that spin in a selected direction, has only two observable values, $\pm\hbar/2$. The corresponding projection of the magnetic moment likewise has only two values.

For an electron, the existence of spin theoretically follows from the relativistic equation of Dirac, but one might consider spin outside special methods of relativistic quantum theory. According to Pauli, spin is an angular momentum, so that it possesses all properties of angular momentum. The eigenvalues of the squared spin angular momentum

$$\mathbf{s}^2 = s_x^2 + s_y^2 + s_z^2$$

are thus $\hbar^2 s(s+1)$; s_x, s_y and s_z are the projections of spin vector $\mathbf{s}$, and s is the spin quantum number. For each elementary particle, the value of s might be defined only from an experiment. For example, $s = 1/2$ for an electron, proton, neutron and μ-meson, $s = 0$ for a π-meson and $s = 1$ for a photon. In the selected representation, one might determine also one projection of spin, for instance, s_z. The possible values for the spin projection number $2s + 1$ in total. For a particle with spin one half, we have two values; these are eigenvalues of variable s_z that equal $\pm\hbar/2$. The classical limit $\hbar \to 0$ yields zero for spin. Classical mechanics fails to explain the presence of the intrinsic angular momentum for these particles; all models involving a spinning top become absurd and yield nothing useful.

To introduce spin into the non-relativistic theory, one must consider the wave equation for the electron in an external magnetic field, with a condition that the electron initially has an intrinsic magnetic moment

$$\boldsymbol{\mu} = -\frac{e}{mc}\mathbf{s},$$

in which c is the speed of light, e is the absolute charge of the electron and m is its mass. We begin from Schrödinger's equation,

$$i\hbar\frac{\partial}{\partial t}|\varphi\rangle = H|\varphi\rangle$$

for states φ; t denotes time. For operator H, we choose the classical expression for a Hamiltonian describing the electron in an external field with vector potential $\mathbf{A}$ and scalar potential U, i.e.

$$H = \frac{1}{2m}\left(\mathbf{p} + \frac{e}{c}\mathbf{A}\right)^2 - eU.$$

Adding to this expression the energy of interaction between the electron magnetic moment and the magnetic field, which is characterized by vector $\mathbf{B}$,

$$W = -\boldsymbol{\mu}\cdot\mathbf{B},$$

we obtain

$$H = \frac{1}{2m}\left(\mathbf{p} + \frac{e}{c}\mathbf{A}\right)^2 - eU - \boldsymbol{\mu}\cdot\mathbf{B}.$$

Thus,

$$i\hbar\frac{\partial}{\partial t}|\varphi\rangle=\left[\frac{1}{2m}\left(\mathbf{p}+\frac{e}{c}\mathbf{A}\right)^2-eU+\frac{e}{mc}\mathbf{s}\cdot\mathbf{B}\right]|\varphi\rangle.$$

Pauli obtained this equation, which describes a motion of the electron in an external electromagnetic field. Pauli's equation is readily generalized for the case of another elementary particle that possesses non-zero spin.

We consider in detail the case $s = 1/2$. With **s**, it is here convenient to introduce a new quantity, $\boldsymbol{\sigma}(\sigma_x, \sigma_y, \sigma_z)$:

$$\mathbf{s}=\frac{\hbar}{2}\boldsymbol{\sigma};$$

for $\boldsymbol{\sigma}$, we have $\boldsymbol{\sigma}\times\boldsymbol{\sigma}=2\mathrm{i}\boldsymbol{\sigma}$. As s_z has eigenvalues $\pm\hbar/2$, component σ_z possesses values ± 1 and σ_z^2 has only one value, +1. Thus,

$$\sigma_x^2=\sigma_y^2=\sigma_z^2=1.$$

Using this equality, we find

$$[\sigma_y^2,\sigma_z]=[1,\sigma_z]=0.$$

Also,

$$[\sigma_y^2,\sigma_z]=\sigma_y[\sigma_y,\sigma_z]+[\sigma_y,\sigma_z]\sigma_y,$$

but $[\sigma_y,\sigma_z]=2\mathrm{i}\sigma_x$, such that

$$\sigma_y\sigma_x+\sigma_x\sigma_y=0 \text{ or } \sigma_y\sigma_x=-\sigma_x\sigma_y.$$

Hence, σ_x and σ_y commute with an opposite sign, i.e. they anticommute. The same conclusions occur for other variables:

$$\begin{aligned}\sigma_x\sigma_y&=-\sigma_y\sigma_x=\mathrm{i}\sigma_z,\\ \sigma_z\sigma_x&=-\sigma_x\sigma_z=\mathrm{i}\sigma_y,\\ \sigma_y\sigma_z&=-\sigma_z\sigma_y=\mathrm{i}\sigma_x.\end{aligned}$$

To determine an explicit form $\boldsymbol{\sigma}$, we recall the formulae obtained earlier for the non-zero matrix elements of raising operator L_+ and lowering operator L_- of angular momentum $\mathbf{L}(L_x, L_y, L_z)$:

$$\langle\ell,k\pm1|L_\pm|\ell k\rangle=\hbar\sqrt{(\ell\mp k)(\ell\pm k+1)}.$$

Here, ℓ is a quantum number that characterizes squared angular momentum $\mathbf{L}^2$ and k correspondingly for projection L_z. As $L_x = (L_+ + L_-)/2$ and $L_y = (L_+ - L_-)/2\mathrm{i}$, then

$$\langle \ell, k+1|L_x|\ell k\rangle = \frac{1}{2}\langle \ell, k+1|L_+|\ell k\rangle = \frac{\hbar}{2}\sqrt{(\ell-k)(\ell+k+1)}$$

and

$$\langle \ell, k+1|L_y|\ell k\rangle = \frac{1}{2\mathrm{i}}\langle \ell, k+1|L_+|\ell k\rangle = -\frac{\mathrm{i}\hbar}{2}\sqrt{(\ell-k)(\ell+k+1)},$$

in which $\langle \ell,k+1|L_-|\ell k\rangle = 0$. In an analogous manner,

$$\langle \ell k|L_x|\ell, k+1\rangle = \frac{1}{2}\langle \ell k|L_-|\ell, k+1\rangle = \frac{\hbar}{2}\sqrt{(\ell-k)(\ell+k+1)}$$

and

$$\langle \ell k|L_y|\ell, k+1\rangle = -\frac{1}{2\mathrm{i}}\langle \ell k|L_-|\ell, k+1\rangle = \frac{\mathrm{i}\hbar}{2}\sqrt{(\ell-k)(\ell+k+1)}.$$

We apply these formulae to spin one half. Assume that $\mathbf{L} = \hbar\boldsymbol{\sigma}/2$, $\ell = s$, and let quantity k retain the preceding meaning of the quantum number of the z-component of angular momentum. We have

$$\langle s, k+1|\sigma_x|sk\rangle = \langle sk|\sigma_x|s, k+1\rangle = \sqrt{(s-k)(s+k+1)}$$

and

$$\langle s, k+1|\sigma_y|sk\rangle = -\langle sk|\sigma_y|s, k+1\rangle = -\,\mathrm{i}\sqrt{(s-k)(s+k+1)};$$

moreover,

$$\langle sk|s_z|sk\rangle = \hbar k \ \text{ and } \ \langle sk|\sigma_z|sk\rangle = 2k;$$

$s = 1/2$, whereas $k = \pm 1/2$; one might consequently represent the components of quantity $\boldsymbol{\sigma}$ in a form of 2×2 Pauli matrices,

$$\sigma_x = \begin{pmatrix} 0 & 1 \\ 1 & 0 \end{pmatrix}, \quad \sigma_y = \begin{pmatrix} 0 & -\mathrm{i} \\ \mathrm{i} & 0 \end{pmatrix}, \quad \sigma_z = \begin{pmatrix} 1 & 0 \\ 0 & -1 \end{pmatrix},$$

$$\sigma_x^2 = \sigma_y^2 = \sigma_z^2 = \begin{pmatrix} 1 & 0 \\ 0 & 1 \end{pmatrix}.$$

The spin variables separately commute with coordinates x, y and z, and also with the components of momentum. For a particle with spin half, the commuting variables (for instance, in the coordinate representation) in a complete set become therefore

$$x, y, z \text{ and } \sigma_z.$$

As σ_z has only two values, ± 1, instead of one-component wave function $\langle xyz\sigma_z|\varphi\rangle$, it is convenient to apply a two-component vector,

$$\begin{pmatrix} \langle xyz, +1|\varphi\rangle \\ \langle xyz, -1|\varphi\rangle \end{pmatrix},$$

which is called a spinor. A spinor is hence a function of three, not four, variables.

We proceed to consider the operator of total angular momentum

$$\mathbf{J} = \mathbf{L} + \mathbf{s},$$
$$J_x = L_x + s_x, \quad J_y = L_y + s_y, \quad J_z = L_z + s_z.$$

As orbital angular momentum $\mathbf{L}$ acts on space coordinates, and $\mathbf{s}$ on spin variables, one might satisfy commutative relations

$$[L_i, s_f] = 0, \ [\mathbf{J}^2, \mathbf{L}^2] = 0 \text{ and } [\mathbf{J}^2, \mathbf{s}^2] = 0,$$

in which i and f can equal x or y or z. Quantity $\mathbf{J}$ retains the general properties that exist for angular momentum, hence

$$\mathbf{J} \times \mathbf{J} = \mathrm{i}\hbar\mathbf{J}$$

and

$$[J_x, \mathbf{J}^2] = [J_y, \mathbf{J}^2] = [J_z, \mathbf{J}^2] = 0.$$

The eigenvalues of J_z, by definition, equal $\hbar k_j$, and of $\mathbf{J}^2$ equal $\hbar^2 j(j+1)$. Number j is expressible through orbital and spin quantum numbers ℓ and s:

$$j = |\ell - s|, \quad |\ell - s| + 1, \ldots, \ell + s - 1, \ell + s.$$

For instance, if $s = 1/2$, then $j = 1/2, 3/2, 5/2, \ldots$ and $k_j = \pm 1/2, \ \pm 3/2, \ldots, \ \pm j$.

The values for the z-projection of $\mathbf{J}$ are obtainable directly through the addition of L_z and s_z, such that

$$k_j = k_\ell + k_s,$$

in which quantum number k_ℓ corresponds to the orbital angular momentum with $2\ell+1$ values and k_s to the spin with $2s+1$ values. For given values of ℓ and s, there must be in total

$$(2\ell+1)(2s+1)$$

various states. The maximally possible value of k_j equals $\ell+s$; only one state corresponds to this value. The maximum of j is hence also equal to $\ell+s$. Decreasing k_j by unity, we obtain $k_j=\ell+s-1$ and two states

$$k_\ell=\ell,\ \ k_s=s-1 \ \text{ and } \ k_\ell=\ell-1,\ \ k_s=s$$

that correspond to this value. Number j has two values, $j=\ell+s$ and $j=\ell+s-1$ at $k_j=\ell+\mathrm{s}-1$. Continuing this scenario with a condition that $s\leq\ell$, we arrive at the value

$$k_j=\ell-s$$

with states of total number $2s+1$. The minimum of j is thus equal to $\ell-s$. According to a classical point of view, in this case, the vectors **L** and **s** are antiparallel to each other, whereas the maximum value $\ell+s$ corresponds to a parallel orientation of angular momenta **L** and **s**. Note that if we continued to decrease k_j by unity, we could not obtain new states; as before, their total number at given ℓ and s equals

$$\sum_{j=\ell-s}^{\ell+s}(2j+1)=(2\ell+1)(2s+1).$$

Semiclassical Theory of Radiation

The transitions of a quantum system induced between particular stationary states attract physical interest. The case of an interaction with an external electromagnetic field is especially important; through this interaction, a system emitting or absorbing a quantum of radiation transfers from one stationary state to another. Not all transitions are, however, allowable; some are weakly probable. The problem of determining the possible transitions and elucidating the corresponding features of the intensity distribution thus arises. According to a semiclassical method, we consider a system to be quantum but a field of radiation to remain classical.

Fermi's Golden Rule

Let a perturbation convey a system from one stationary state to another under a condition that the states remain unaltered; the problem then becomes non-stationary.

To define the probability of such a transition for a quantum system, our point of departure is Schrödinger's equation,

$$i\hbar\frac{\partial}{\partial t}|\varphi\rangle = (H^0 + \lambda W)|\varphi\rangle$$

for states φ, in which H^0 is the Hamiltonian of zero order with eigenvalues E_n and eigenfunctions

$$e^{-iE_nt/\hbar}|n\rangle,$$

λ is a small parameter characterizing the order of perturbation W and t denotes time. Having expanded $|\varphi\rangle$ in vectors of the unperturbed Hamiltonian,

$$|\varphi\rangle = \sum_n a_n(t)e^{-iE_nt/\hbar}|n\rangle,$$

in which a_n are the amplitudes of states, we proceed to the equation

$$i\hbar\frac{\partial a_m}{\partial t} = \lambda\sum_n a_n\, e^{i\omega_{mn}t}\langle m|W|n\rangle,\ \omega_{mn} = \frac{E_m - E_n}{\hbar}.$$

Here, we take into account that $H^0|n\rangle = E_n|n\rangle$ and $\langle m|n\rangle = \delta_{mn}$.

We represent a_n in a form of an expansion in the small parameter,

$$a_n = a_n^0 + \lambda a_n^1 + \cdots,$$

substitute this expansion into the equation for amplitudes a_n and restrict our consideration to the first order in λ, then

$$i\hbar\frac{\partial a_m^1}{\partial t} = \sum_n a_n^0\, e^{i\omega_{mn}t}\langle m|W|n\rangle.$$

Taking into account that an initial state, for instance,

$$e^{-iE_kt/\hbar}|k\rangle$$

is determined, i.e. $a_n^0 = \delta_{nk}$, we define a_m^1. We have

$$a_m^1 = \frac{1}{i\hbar}\int_0^t \langle m|W|k\rangle e^{i\omega_{mk}t}\, dt.$$

If perturbation W is independent of time,

$$a_m^1 = -\frac{e^{i\omega_{mk}t} - 1}{\hbar\omega_{mk}} \langle m|W|k\rangle.$$

One might determine the probability of transition per time t as

$$\rho_{mk} = \frac{|a_m^1|^2}{t} = \frac{4\sin^2(\omega_{mk}t/2)}{t\hbar^2\omega_{mk}^2} |\langle m|W|k\rangle|^2.$$

At a sufficiently large value of t, one obtains

$$\rho_{mk} = \frac{2\pi}{\hbar^2} |\langle m|W|k\rangle|^2 \delta(\omega_{mk}),$$

in which we take into account this representation for Dirac's delta function,

$$\delta(\omega_{mk}) = \lim_{t'\to\infty} \frac{\sin^2(t'\omega_{mk})}{\pi t'\omega_{mk}^2}.$$

The sought probability of a transition from state k to state m per unit time thus equals

$$\rho_{mk} = \frac{2\pi}{\hbar} |\langle m|W|k\rangle|^2 \delta(E_m - E_k).$$

We see that, for first-order transitions under an influence of a perturbation, the law of conservation of energy is satisfied. The formula obtained is called Fermi's golden rule. If a transition occurs from some discrete state to a state belonging to a continuous spectrum, instead of $\delta(E_m - E_k)$ one must apply the density of finite states $\wp(E_m)$.

Intensities of Transitions

We consider an arbitrary physical system with electric dipolar moment **d** in an external electric field with field vector **E**. Let

$$H = H^0 + W$$

be the exact Hamiltonian of the system, and H^0 be the zero-order Hamiltonian; H^0 possesses eigenvalues E_n and eigenfunctions $|n\rangle$. The energy of interaction

$$W = -\mathbf{d} \cdot \mathbf{E}$$

plays a role of a perturbation and causes the transitions of the system between various states of zero-order approximation. The sources of field are lacking; field vector $\mathbf{E}$ satisfying a classical wave equation has the form of a plane wave,

$$\mathbf{E} = \mathbf{E}_0 e^{i(\mathbf{k}\cdot\mathbf{r} - \omega t)},$$

in which $\mathbf{E}_0$ is the constant complex vector, $\mathbf{r}$ is the radius vector of the current space point, ω is the frequency of the field and $\mathbf{k}$ is the wave vector; if c is the speed of light, $c|\mathbf{k}| = \omega$. Our purpose is to calculate the intensity of the transition of a system from state $|n\rangle$ to state $|m\rangle$.

The real part of the field vector

$$\mathrm{Re}\,\mathbf{E} = \frac{1}{2}(\mathbf{E} + \mathbf{E}^*)$$

hence represents our physical interest, namely that $\mathrm{Re}\,\mathbf{E}$ must be substituted into W. One should, furthermore, substitute W into the expression obtained above for the transition probability per unit time. We have

$$W = \left(-\frac{\mathbf{d}\cdot\mathbf{E}_0}{2} e^{i\mathbf{k}\cdot\mathbf{r}}\right) e^{-i\omega t} + \left(-\frac{\mathbf{d}\cdot\mathbf{E}_0^*}{2} e^{-i\mathbf{k}\cdot\mathbf{r}}\right) e^{i\omega t};$$

according to Fermi's golden rule, for absorption,

$$\rho_{n\to m} = \frac{2\pi}{\hbar^2}\left|\left\langle n\left|\frac{\mathbf{d}\cdot\mathbf{E}_0}{2} e^{i\mathbf{k}\cdot\mathbf{r}}\right|m\right\rangle\right|^2 \delta(\omega_{mn} - \omega)$$

and, for emission,

$$\rho_{n\to m} = \frac{2\pi}{\hbar^2}\left|\left\langle n\left|\frac{\mathbf{d}\cdot\mathbf{E}_0^*}{2} e^{-i\mathbf{k}\cdot\mathbf{r}}\right|m\right\rangle\right|^2 \delta(\omega_{mn} + \omega).$$

Combining these formulae, one obtains

$$\rho_{n\to m} = \frac{\pi}{2}\left(\frac{E_0}{\hbar}\cos\theta\right)^2 |\langle n|e^{\pm i\mathbf{k}\cdot\mathbf{r}}\mathbf{d}|m\rangle|^2 \delta(\omega_{mn} \mp \omega), \qquad \omega_{mn} = \frac{E_m - E_n}{\hbar},$$

in which the upper sign corresponds to an absorption and the lower sign to an emission; θ is the angle between vectors $\mathbf{d}$ and $\mathbf{E}_0$. We average this expression with respect to all possible values of angle θ; this procedure yields the factor

$$\frac{1}{4\pi}\int_{4\pi} \cos^2\theta \, d\Omega = \frac{1}{4\pi}\int_0^{\pi} \cos^2\theta \cdot 2\pi \sin\theta \, d\theta = \frac{1}{3},$$

in which Ω is the solid angle. As a result,

$$\rho_{n\to m} = \frac{\pi}{6}\left(\frac{E_0}{\hbar}\right)^2 |\langle n|e^{\pm i\mathbf{k}\cdot\mathbf{r}}\mathbf{d}|m\rangle|^2\delta(\omega_{mn} \mp \omega).$$

One might relate the squared amplitude E_0^2 of the electric field to the number of incident quanta per unit time and per unit square $C(\omega)$. This relation allows one to proceed from a semiclassical consideration to the quantum domain. Quantity $\hbar\,\Delta\omega C(\omega)$ that represents the intensity of incident radiation per unit square in a small frequency interval $\Delta\omega$ is the expectation value of the Poynting vector,

$$\frac{c}{4\pi}\langle(\mathrm{Re}\,\mathbf{E})^2\rangle;$$

the averaging is here performed according to this formula,

$$\langle(\mathrm{Re}\,\mathbf{E})^2\rangle = \lim_{t'\to\infty}\frac{1}{t'}\int_0^{t'}(\mathrm{Re}\,\mathbf{E})^2\,\mathrm{d}t.$$

Noticing that the expectation values of products **EE** and $\mathbf{E}^*\mathbf{E}^*$ equal zero, we have

$$\langle(\mathrm{Re}\mathbf{E})^2\rangle = E_0^2/2.$$

Consequently,

$$\hbar\,\Delta\omega\,C(\omega) = \frac{cE_0^2}{8\pi}.$$

Eliminating E_0^2 with the aid of this result, we find

$$\rho_{n\to m} = \frac{4\pi^2}{3c\hbar}C(\omega)|\langle n|e^{\pm i\mathbf{k}\cdot\mathbf{r}}\mathbf{d}|m\rangle|^2\delta(\omega_{mn}\mp\omega)\Delta\omega.$$

Summing $\rho_{n\to m}$ over all frequencies corresponds to integration with respect to ω with a condition of non-coherent addition of transition probabilities, which are stipulated by incident waves of various frequencies. As a result,

$$\rho_{n\to m} = \frac{4\pi^2}{3c\hbar}C(\pm\omega_{mn})|\langle n|e^{\pm i\mathbf{k}\cdot\mathbf{r}}\mathbf{d}|m\rangle|^2.$$

As above, the plus sign corresponds to absorption and minus to emission. For the emission, the non-negativity of the argument in C indicates that a system makes a transition from a more excited state to one less excited,

$$-\omega_{mn} = -\frac{E_m - E_n}{\hbar} = \frac{E_n - E_m}{\hbar} = \omega_{nm} > 0.$$

There remains to multiply $\rho_{n\to m}$ by the transition energy $\hbar\omega_{mn}$, so to obtain the sought expression for intensity

$$I_{n\to m} = \frac{4\pi^2}{3c}(\pm\omega_{mn})C(\pm\,\omega_{mn})|\langle n|e^{\pm\,\mathrm{i}\mathbf{k}\cdot\mathbf{r}}\mathbf{d}|m\rangle|^2.$$

In such a form, this formula is rarely applicable. For many physical problems, it is convenient to invoke a further simplification. One might note that the argument of the exponential function becomes a sufficiently small quantity,

$$\mathbf{k}\cdot\mathbf{r}\sim\frac{\omega}{c}a_0 = \frac{\Delta E}{\hbar}\frac{a_0}{c}\sim\frac{e^2}{2a_0}\frac{a_0}{\hbar c} = \frac{e^2}{2\hbar c}\approx\frac{1}{300},$$

for which we suppose that energy ΔE of an optical transition is at most one rydberg that equals $e^2/2a_0$; here a_0 is the Bohr radius and e is the absolute value of charge of an electron. One might hence expand the exponential function into this series,

$$\mathrm{e}^{\pm\,\mathrm{i}\mathbf{k}\cdot\mathbf{r}} = 1\pm\mathrm{i}\mathbf{k}\cdot\mathbf{r}+\cdots$$

in powers of $\mathbf{k}\cdot\mathbf{r}$ and restrict our attention to unity, then

$$I_{n\to m} = \frac{4\pi^2}{3c}(\pm\omega_{mn})C(\pm\,\omega_{mn})|\langle n|\mathbf{d}|m\rangle|^2.$$

This result represents an application of the electric-dipole approximation: the intensity of the transition is determined by the matrix element of the function for the electric dipolar moment of a system. If

$$\langle n|\mathbf{d}|m\rangle = 0,$$

the transition $n\to m$ is forbidden, but only in the electric-dipole approximation. In this case, in an expansion of the exponential function one must retain the next term $\pm\mathrm{i}\mathbf{k}\cdot\mathbf{r}$; the intensity is then expressible through matrix element

$$\langle n|(\mathbf{k}\cdot\mathbf{r})\mathbf{d}|m\rangle.$$

Such an approximation corresponds to electric quadrupolar and magnetic dipolar transitions in combination. In an analogous manner, the matrix elements of higher order are readily obtainable; they correspond to higher multipole transitions. If

$$\langle n|e^{\pm\,\mathrm{i}\mathbf{k}\cdot\mathbf{r}}\mathbf{d}|m\rangle = 0,$$

the transition $n\to m$ is strictly forbidden and to evaluate the intensity, one must consider the second order of perturbation theory.

Apart from these induced transitions, a spontaneous alteration of a state of a system might occur: we here imply a spontaneous emission – there is no such process for absorption. One might obtain the intensity of spontaneous emission through either a consistent calculation in the framework of quantum electrodynamics or a phenomenological method involving Einstein's probabilities. We focus attention on the second method.

We consider a system of atoms at thermodynamic equilibrium and a radiation field at temperature T for transitions $n \leftrightarrow m$ of only one type. For the equilibrium emission, we introduce a density function $u(\omega,T)$, which equals the energy of emission per unit volume, for instance 1 cm^3 in a frequency range $[\omega,\omega + \mathrm{d}\omega]$. Restricting ourselves to a dipole approximation, we rewrite the obtained semiclassical expression for ρ:

$$\rho_{n \leftrightarrow m} = \frac{\pi}{6\hbar^2} E_0^2 |\langle n|\mathbf{d}|m\rangle|^2 \delta(\omega_{mn} \mp \omega).$$

One sees that $E_0^2/8\pi$ clearly represents the energy density – energy per unit volume; thus, $E_0^2/8\pi = u(\omega, T)\mathrm{d}\omega$ and, after integration with respect to ω,

$$\rho_{n \leftrightarrow m} = \frac{4\pi^2}{3\hbar^2} u(\omega_{mn}, T) |\langle n|\mathbf{d}|m\rangle|^2.$$

This expression defines the probability of an induced transition per unit time for absorption $n \rightarrow m$ and emission $n \leftarrow m$; $\rho_{n \leftrightarrow m}$ is also called the rate of the transition. Multiplying ρ by the number of atoms in initial state N, we determine the total rate of absorption

$$\frac{\mathrm{d}N(n \rightarrow m)}{\mathrm{d}t} = N(n) B_{mn} u(\omega_{mn}, T),$$

$$B_{mn} = \frac{4\pi^2}{3\hbar^2} |\langle n|\mathbf{d}|m\rangle|^2.$$

For emission, analogously

$$\frac{\mathrm{d}N(n \leftarrow m)}{\mathrm{d}t} = N(m) B_{mn} u(\omega_{mn}, T) + A_{mn} N(m),$$

in which, according to Einstein, we take into account the probability of spontaneous emission A_{mn}.

For thermal equilibrium,

$$\frac{\mathrm{d}N(n \rightarrow m)}{\mathrm{d}t} = \frac{\mathrm{d}N(n \leftarrow m)}{\mathrm{d}t},$$

where from

$$u(\omega_{mn}, T) = \frac{A_{mn}/B_{mn}}{(N(n)/N(m)) - 1}.$$

Through the canonical distribution,

$$N(n) \sim \exp(-E_n/k_B T) \text{ and } N(m) \sim \exp(-E_m/k_B T),$$

in which k_B is the Boltzmann constant, hence

$$u(\omega_{mn}, T) = \frac{A_{mn}}{B_{mn}} \frac{1}{e^{\hbar\omega_{mn}/k_B T} - 1}.$$

This formula must become converted into empirical expressions of Wien,

$$u(\omega_{mn}, T) = \frac{\hbar\omega_{mn}^3}{\pi^2 c^3} e^{-\hbar\omega_{mn}/k_B T} \text{ at } \hbar\omega_{mn} \gg k_B T$$

and of Rayleigh–Jeans,

$$u(\omega_{mn}, T) = k_B T \frac{\omega_{mn}^2}{\pi^2 c^3} \text{ at } \hbar\omega_{mn} \ll k_B T.$$

Thus,

$$A_{mn} = \frac{\hbar\omega_{mn}^3}{\pi^2 c^3} B_{mn} = \frac{4\omega_{mn}^3}{3\hbar c^3} |\langle n|\mathbf{d}|m\rangle|^2$$

and

$$u(\omega_{mn}, T) = \frac{\hbar\omega_{mn}^3}{\pi^2 c^3} \frac{1}{e^{\hbar\omega_{mn}/k_B T} - 1}.$$

This expression for the energy density represents Planck's well-known formula for the radiation from a black body.

We elucidate the physical meaning of quantity A_{mn}. Let

$$dN(m) = -A_{mn} N(m) dt$$

be the number of spontaneous transitions per time dt; the minus sign indicates a decrease of the total number of atoms from an initial state. We calculate the average lifetime of an atom in an excited state. We have

$$\langle t \rangle = \int_0^\infty t \frac{|dN(m)|}{N_0(m)},$$

in which $N_0(m)$ denotes the value of $N(m)$ at $t = 0$;

$$N(m) \sim e^{-A_{mn}t}.$$

Hence,

$$\langle t \rangle = A_{mn} \int_0^\infty t\, e^{-A_{mn}t}\, dt = \frac{1}{A_{mn}} = \frac{3\hbar c^3}{4\omega_{mn}^3} |\langle n|\mathbf{d}|m\rangle|^{-2}$$

and coefficient A_{mn} implies the duration of excited state $|m\rangle$. Moreover, as

$$\langle t \rangle \neq 0,$$

we obtain some indeterminacy in the energy of the state under consideration. The spectral lines thus fail to become indefinitely narrow and possess a natural finite width provided by time $\langle t \rangle$. This circumstance is sometimes expressed in a form of the relation of indeterminacy,

$$\Delta E\, \Delta t \sim \hbar,$$

which relates the indeterminacy in energy of some state ΔE to its lifetime Δt.

Second Quantization

We consider a quantum-mechanical description of systems comprising multiple identical particles. That identity emphasizes that the physical properties of the particles, the role of which, in a context of a concrete problem, might be played by photons, electrons, atoms and even molecules, are indistinguishable. One might substitute one particle for another, and the state of the system remains invariant. For instance, let H be the Hamiltonian of a system comprising two identical particles. We write the equation for eigenvalues E and eigenfunctions $|1,2\rangle$ in a form

$$H|1,2\rangle = E|1,2\rangle;$$

by numbers 1 and 2 we imply all variables corresponding to the first and second particles, respectively. Through the identity of particles,

$$H|2,1\rangle = E|2,1\rangle.$$

Function $|1,2\rangle$ thus differs from $|2,1\rangle$ by only a constant coefficient. Denoting this coefficient by b, we have

$$|1,2\rangle = b|2,1\rangle = b^2|1,2\rangle,$$

where from $b = \pm 1$ and two variants become possible. If

$$|1,2\rangle = |2,1\rangle,$$

we treat a symmetric state; if

$$|1,2\rangle = -|2,1\rangle,$$

antisymmetric.

This assertion extrapolates to a system of many bodies. To show it, we introduce operator P_{ij} that interchanges particles i and j:

$$P_{ij}|1,2,\ldots,i,\ldots,j,\ldots\rangle = |1,2,\ldots,j,\ldots,i,\ldots\rangle = \pm|1,2,\ldots,i,\ldots,j,\ldots\rangle.$$

We choose a plus sign for symmetric and minus for antisymmetric states. An experiment shows that all wave functions for particles satisfy the relation of either symmetry or antisymmetry. For instance, π-mesons, photons and K-mesons are described by symmetric wave functions, whereas electrons, protons and neutrons are described by antisymmetric ones.

Permutation operator P, commuting with Hamiltonian H, is a constant of the motion. Thus,

$$HP|\cdots\rangle = PH|\cdots\rangle$$

and the Hamiltonian does not alter the symmetry of an initial state:

$$H|\text{state}\rangle = |\text{state of the same symmetry}\rangle.$$

From another point of view, according to Schrödinger's equation

$$H|\cdots\rangle = \mathrm{i}\hbar\partial|\cdots\rangle/\partial t,$$

the derivative of the state vector with respect to time t hence also maintains its initial symmetry. As

$$\partial|\cdots\rangle/\partial t$$

defines the state vector at an arbitrary moment in time, physical states possessing a particular symmetry maintain it with time. If, at an initial moment in time, the symmetry of some state is symmetric or antisymmetric, it remains so forever.

We consider the symmetric and antisymmetric states in detail. We introduce the Hamiltonian of the system comprising k particles,

$$H = H_1 + H_2 + \cdots + H_k;$$

it represents a sum of one-particle Hamiltonians H_i, $i = 1, 2,\ldots, k$. Let $\Phi_{n_i}(i)$ and E_{n_i} be the set of eigenfunctions and eigenvalues of H_i, then

$$\Phi = \Phi_{n_1}(1)\Phi_{n_2}(2)\cdots\Phi_{n_k}(k)$$

is a possible eigenfunction of Hamiltonian H with eigenvalue

$$E = E_{n_1} + E_{n_2} + \cdots + E_{n_k},$$

in which n_i designates states that are accessible to the particles. For electrons, the one-particle wave functions Φ_{n_i} are commonly called orbitals.

Interchanging the particles randomly for each other, i.e. a random permutation of numbers in parentheses of functions

$$\Phi_{n_1}(1),\ \Phi_{n_2}(2), \ldots,\ \Phi_{n_k}(k),$$

we retain the value of E to be invariant. As there are $k!$ permutations of these identical particles, the value E becomes degenerate with degeneracy $k!$; $k!$ various functions correspond to this value. Such degeneracy is called exchange.

One might note, however, that chosen function Φ as a form of product of the one-particle wave functions fails to be either symmetric or antisymmetric. How can we make our function symmetric or antisymmetric? We form a set of wave functions through all possible permutations of subscripts n_i; the sum of all obtained functions,

$$\sum_P \Phi_{P(n_1)}(1)\Phi_{P(n_2)}(2)\cdots\Phi_{P(n_k)}(k)$$

yields a symmetric state, whereas the sum

$$\sum_P (-1)^P \Phi_{P(n_1)}(1)\Phi_{P(n_2)}(2)\cdots\Phi_{P(n_k)}(k)$$

describes an antisymmetric state; $(-1)^P = -1$ for odd permutations and $(-1)^P = 1$ for even.

The normalized antisymmetric wave function is represented in a form called a Slater determinant,

$$\frac{1}{\sqrt{k!}}\begin{vmatrix} \Phi_{n_1}(1) & \Phi_{n_1}(2) & \cdots & \Phi_{n_1}(k) \\ \Phi_{n_2}(1) & \Phi_{n_2}(2) & \cdots & \Phi_{n_2}(k) \\ \vdots & \vdots & \ddots & \vdots \\ \Phi_{n_k}(1) & \Phi_{n_k}(2) & \cdots & \Phi_{n_k}(k) \end{vmatrix}.$$

On interchanging the particles or their states, the sign of the determinant becomes automatically reversed, such that this function describes a truly antisymmetric state.

A determinant with two identical rows is inevitably equal to zero; two particles, for instance, two electrons, can therefore not occupy one and the same state. This fact leads to Pauli's exclusion principle: in one system, only one electron might be in one state. Pauli's principle allows one to explain Mendeleev's periodic table of the chemical elements.

We generalize the above facts. The symmetric wave functions correspond to particles with integer spin, and the antisymmetric functions accord to particles with half-integer spin. Particles of the former type conform to Bose–Einstein statistics; such particles are called bosons. The latter type is called a fermion; it conforms to Fermi–Dirac statistics. A complicated physical system includes both bosons and fermions. Such a system (for instance, an atom or a molecule) is a boson if it has an even number of fermions. A complicated system containing an odd number of fermions is a fermion.

A Harmonic Oscillator

In quantum mechanics, the method of creation and destruction operators is widely used to describe a many-body system. This approach is a transformation to an energy representation, for which one might, indirectly, think that energy becomes the pertinent variable. This method, based on a simple operator problem, we consider in detail. Let a be some operator that satisfies this commutation relation,

$$[a, a^+] = aa^+ - a^+a = 1,$$

in which a^+ is an operator Hermitian conjugate to a. Our main purpose is to define eigenvalues φ of the product a^+a.

As $[a,a] = 0$ and $[a^+,a^+] = 0$, these equalities are satisfied:

$$a = [a, a^+]a = -[a^+a, a] \text{ and } a^+ = a^+[a, a^+] = [a^+a, a^+].$$

Consequently,

$$(a^+a)a = a(a^+a - 1) \text{ and } (a^+a)a^+ = a^+(a^+a + 1).$$

Applying the latter relations, we find

$$(a^+a)a|\varphi\rangle = a(a^+a - 1)|\varphi\rangle = a(\varphi - 1)|\varphi\rangle = (\varphi - 1)a|\varphi\rangle,$$

$$(a^+a)a^+|\varphi\rangle = a^+(a^+a + 1)|\varphi\rangle = a^+(\varphi + 1)|\varphi\rangle = (\varphi + 1)a^+|\varphi\rangle,$$

in which $|\varphi\rangle$ is the normalized eigenvector of operator a^+a;

$$a^+a|\varphi\rangle = \varphi|\varphi\rangle.$$

Thus, if

$$a|\varphi\rangle \neq 0 \text{ and } a^+|\varphi\rangle \neq 0,$$

$a|\varphi\rangle$ and $a^+|\varphi\rangle$ are eigenvectors of operator a^+a with eigenvalues $\varphi - 1$ and $\varphi + 1$, respectively. The relation

$$||a|\varphi\rangle||^2 = \langle\varphi|a^+a|\varphi\rangle,$$

in which $a^+a|\varphi\rangle = \varphi|\varphi\rangle$ with $\langle\varphi|\varphi\rangle = 1$ defines the norm of vector $a|\varphi\rangle$;

$$||a|\varphi\rangle|| = \sqrt{\varphi}.$$

Analogously, through $aa^+|\varphi\rangle = (a^+a + 1)|\varphi\rangle = (\varphi + 1)|\varphi\rangle$, we find

$$||a^+|\varphi\rangle|| = \sqrt{\varphi + 1}.$$

Furthermore, let n be a non-negative integer, then

$$\begin{aligned}(a^+a)a^n|\varphi\rangle &= a(a^+a - 1)a^{n-1}|\varphi\rangle = a[(a^+a)a^{n-1} - a^{n-1}]|\varphi\rangle \\ &= a[a(a^+a)a^{n-2} - 2a^{n-1}]|\varphi\rangle = \cdots = (\varphi - n)a^n|\varphi\rangle,\end{aligned}$$

and also

$$(a^+a)(a^+)^n|\varphi\rangle = (\varphi + n)(a^+)^n|\varphi\rangle;$$

$a^n|\varphi\rangle$ and $(a^+)^n|\varphi\rangle$ are hence eigenvectors of product a^+a with eigenvalues $\varphi - n$ and $\varphi + n$. Vector $|\varphi - n\rangle$ belongs to value $\varphi - n$, such that

$$\frac{a^n|\varphi\rangle}{||a^n|\varphi\rangle||} = |\varphi - n\rangle.$$

However, $||a|\varphi - n\rangle|| = \sqrt{\varphi - n}$; thus, $\varphi \geq n$. There must hence exist number n for which

$$a|\varphi - n\rangle = \frac{a^{n+1}|\varphi\rangle}{||a^n|\varphi\rangle||} = 0$$

and $||a|\varphi - n\rangle|| = \sqrt{\varphi - n} = 0$, where from

$$\varphi = n.$$

What do we see? The sought eigenvalues of operator a^+a are positive integers. Moreover, through this equation

$$a|0\rangle = 0,$$

one defines the ground state $|0\rangle$. These conclusions are mutually related. Having determined the ground-state vector, we may act on it n times with operator a^+; then through relation

$$(a^+a)(a^+)^n|\varphi\rangle = (\varphi + n)(a^+)^n|\varphi\rangle,$$

we obtain

$$(a^+a)(a^+)^n|0\rangle = n(a^+)^n|0\rangle;$$

non-negative integers 0, 1, 2,... hence play a role as the eigenvalues of quantity a^+a.

Assuming $\varphi = n$ in these expressions

$$a^+a|\varphi\rangle = \varphi|\varphi\rangle,$$

$$\frac{a|\varphi\rangle}{||a|\varphi\rangle||} = |\varphi - 1\rangle \text{ and } \frac{a^+|\varphi\rangle}{||a^+|\varphi\rangle||} = |\varphi + 1\rangle,$$

we eventually obtain

$$a^+a|n\rangle = n|n\rangle, \quad n = 0,\ 1,\ 2, \ldots,$$

$$a|n\rangle = \sqrt{n}|n-1\rangle \text{ and } a^+|n\rangle = \sqrt{n+1}|n+1\rangle.$$

These relations constitute the solution of the problem in question. With their aid, one might readily find general expressions for the eigenvectors. We have

$$a|0\rangle = 0,$$

$$|1\rangle = a^+|0\rangle, |2\rangle = \frac{1}{\sqrt{2}}a^+|1\rangle = \frac{(a^+)^2|0\rangle}{\sqrt{2!}}, \ldots, |n\rangle = \frac{(a^+)^n|0\rangle}{\sqrt{n!}}.$$

These vectors are orthonormal:

$$\langle n|m\rangle = \frac{1}{\sqrt{n!m!}}\langle 0|a^n(a^+)^m|0\rangle = \delta_{nm}.$$

Quantity a^+a is called a number operator; a^+ and a are correspondingly the operators of creation and destruction. The creation operator increases, whereas the destruction operator decreases, the number n characterizing the state of operator a^+a by unity. In many-body physics, it is necessary to work not only with particles but also with excitations, which are indirectly associated with particles, but the latter become virtual. For particles of each kind, in this case, one might define their own pair of creation and destruction operators. Treating excitations, one should understand that their occupation numbers represent the energy of interaction between physical objects. The transformation from natural variables for classical physics to the operators of creation and destruction therefore corresponds to a transformation into the energy representation.

As an example of the method just developed, we consider a harmonic oscillator. This example is of importance to understand physical processes, which are concerned with atomic and molecular vibrations, the theory of radiation, aspects of quantum-field theory and many other questions. The Hamiltonian of a one-dimensional harmonic oscillator is given in a form

$$H = \frac{p^2}{2m} + \frac{m\omega^2 x^2}{2},$$

in which appear mass m, momentum p and displacement x from an equilibrium point of a particle that makes small vibrations with frequency ω; quantities x and p satisfy the commutation relation

$$[x, p] = \mathrm{i}\hbar.$$

Supposing x and p to be classical variables, we transform Hamiltonian H. We have

$$H_{\text{class}} = \hbar\omega\left(\frac{p^2}{2m\omega\hbar} + \frac{m\omega}{2\hbar}x^2\right) = \hbar\omega\left(\sqrt{\frac{m\omega}{2\hbar}}x - \mathrm{i}\frac{p}{\sqrt{2m\omega\hbar}}\right)\left(\sqrt{\frac{m\omega}{2\hbar}}x + \mathrm{i}\frac{p}{\sqrt{2m\omega\hbar}}\right),$$

in which one should understand H_{class} in a classical meaning such that $[x,p] = 0$. We introduce a new quantity,

$$\eta = \frac{1}{\sqrt{2}}\left(\sqrt{\frac{m\omega}{\hbar}}x + \mathrm{i}\frac{p}{\sqrt{m\omega\hbar}}\right);$$

then

$$H_{\text{class}} = \hbar\omega\eta^*\eta.$$

Let us seek what this classical expression yields in quantum mechanics. Supposing x and p to be operators, we have

$$\eta = \frac{1}{\sqrt{2}}\left(\sqrt{\frac{m\omega}{\hbar}}x + \mathrm{i}\frac{p}{\sqrt{m\omega\hbar}}\right) \text{ and } \eta^+ = \frac{1}{\sqrt{2}}\left(\sqrt{\frac{m\omega}{\hbar}}x - \mathrm{i}\frac{p}{\sqrt{m\omega\hbar}}\right);$$

$$\left[\eta, \eta^+\right] = -\frac{\mathrm{i}}{\hbar}[x, p] = 1.$$

Here, we take into account that $x = x^+$ and $p = p^+$. Furthermore,

$$H_{\text{class}} \to \hbar\omega\eta^+\eta;$$

consequently,

$$\hbar\omega\eta^+\eta = \frac{p^2}{2m} + \frac{m\omega^2x^2}{2} + \frac{\mathrm{i}}{2\hbar}\hbar\omega[x, p] = H - \frac{\hbar\omega}{2},$$

where from

$$H = \hbar\omega\left(\eta^+\eta + \frac{1}{2}\right).$$

As we see, the distinction between H and H_{class} consists of the appearance of an additional constant quantity $\hbar\omega/2$.

To find the energy levels of a harmonic oscillator, one must therefore solve the problem for the eigenvalues of operator $\eta^+\eta$ with the condition that

$$[\eta, \eta^+] = 1.$$

The solution of this problem is already known. Having put, by definition,

$$\eta^+\eta|n\rangle = n|n\rangle, \quad n = 0, 1, 2, \ldots,$$

we obtain

$$H|n\rangle = \hbar\omega\left(\eta^+\eta|n\rangle + \frac{1}{2}|n\rangle\right) = \hbar\omega\left(n + \frac{1}{2}\right)|n\rangle.$$

As

$$H|n\rangle = E_n|n\rangle,$$

consequently,

$$E_n = \hbar\omega\left(n + \frac{1}{2}\right), \quad n = 0, 1, 2, \ldots.$$

These are the sought energy levels of a quantum-mechanical harmonic oscillator. The adjacent levels, as we see, are separate from each other by constant quantity $\hbar\omega$, such that the levels of the oscillator are distributed in an equidistant manner. The least possible value of energy equals $E_0 = \hbar\omega/2$, not zero as in classical mechanics.

Let us construct the eigenvectors for the found values E_n. It is convenient to work in the coordinate representation. In this case,

$$p = -\mathrm{i}\hbar\frac{\partial}{\partial x}, \quad \eta = \frac{1}{\sqrt{2}}\left(\sqrt{\frac{m\omega}{\hbar}}x + \sqrt{\frac{\hbar}{m\omega}}\frac{\partial}{\partial x}\right) \quad \text{and } |n\rangle \rightarrow \varphi_n(x).$$

We find the function for the ground state from this equation,

$$\eta\varphi_0(x) = 0$$

or, in an explicit form,

$$x\varphi_0(x) + \frac{\hbar}{m\omega}\frac{\partial}{\partial x}\varphi_0(x) = 0,$$

hence

$$\varphi_0(x) = C\mathrm{e}^{-m\omega x^2/2\hbar}.$$

Constant C of integration is defined through a normalization condition,

$$\langle 0|0\rangle = |C|^2\int_{-\infty}^{+\infty}\mathrm{e}^{-m\omega x^2/\hbar}\mathrm{d}x = 1;$$

thus $C = (m\omega/\pi\hbar)^{1/4}$. Consequently,

$$\varphi_0(x) = \left(\frac{m\omega}{\pi\hbar}\right)^{1/4}\mathrm{e}^{-m\omega x^2/2\hbar}.$$

We determine the vector of an arbitrary state with the aid of relation

$$|n\rangle = \frac{1}{\sqrt{n!}}(\eta^+)^n|0\rangle.$$

We have

$$\varphi_n(x) = \frac{1}{\sqrt{n!}}\left(\frac{m\omega}{\pi\hbar}\right)^{1/4}\left(\frac{m\omega}{2\hbar}\right)^{n/2}\left(x - \frac{\hbar}{m\omega}\frac{\partial}{\partial x}\right)^n\mathrm{e}^{-m\omega x^2/2\hbar}.$$

The vectors obtained in the coordinate representation are expressible through Chebyshev–Hermite polynomials

$$H_n(\xi) = (-1)^n\,\mathrm{e}^{\xi^2}\frac{\mathrm{d}^n}{\mathrm{d}\xi^n}\mathrm{e}^{-\xi^2},$$

namely

$$\varphi_n(\xi) = A\,e^{-\xi^2/2} H_n(\xi), \quad \xi = x\sqrt{m\omega/\hbar}.$$

Coefficient A is chosen so that function $\varphi_n(\xi)$ becomes normalized to unity; $H_0 = 1$, $H_1 = 2\xi$ and so on.

The Fields of Bosons and Fermions

Up to this point, we have considered a description in terms of the coordinates and momenta; both latter quantities, according to the general ideology of quantum theory, are expressible through quantized values, whereas the fields concerned with these variables remain classical. A correct description requires a revision of the theory. One might achieve this purpose with the aid of second quantization. The occupation numbers of separate particles in particular states become variables after second or repeated quantization. The new formalism constitutes a basis of quantum electrodynamics; with its aid, one might solve the problems of quantization not only of the electromagnetic field but also of the one-particle fields of the Schrödinger, Klein–Fock–Gordon and Dirac equations.

We begin with a consideration of a boson field. Let field operators $\varphi(\mathbf{r})$ and $\varphi^+(\mathbf{r})$ satisfy the commutation relations

$$[\varphi(\mathbf{r}), \varphi^+(\mathbf{r}')] = \delta(\mathbf{r} - \mathbf{r}'),$$

$$[\varphi(\mathbf{r}), \varphi(\mathbf{r}')] = 0 \text{ and } [\varphi^+(\mathbf{r}), \varphi^+(\mathbf{r}')] = 0,$$

in which $\mathbf{r}$ and $\mathbf{r}'$ are the radius vectors of two arbitrary points; all quantities are here taken at one and the same moment in time. Through these terms, the one-particle function of Hamilton for the particle moving in a field with potential $V(\mathbf{r})$ has a form

$$H = \int \varphi^+ \left(-\frac{\hbar^2}{2m} \nabla^2 + V(\mathbf{r}) \right) \varphi \, d\tau,$$

in which m is the mass of the particle and $d\tau$ is an element of volume. One might arrive at this expression through a traditional formalism of analytical mechanics of fields with Lagrange's function density

$$\Im = i\hbar \varphi^+ \dot{\varphi} - \frac{\hbar^2}{2m} (\nabla \varphi^+)(\nabla \varphi) - V \varphi^+ \varphi.$$

We prefer, however, to postulate that a Hamiltonian, not a Lagrangian, is appropriate here.

Let us consider an equation of the motion for $\varphi(\mathbf{r})$; we have

$$\mathrm{i}\hbar\dot{\varphi}=[\varphi,H]=\left[\varphi(\mathbf{r}),\int\varphi^{+}(\mathbf{r}')\left(-\frac{\hbar^2}{2m}\nabla'^2\right)\varphi(\mathbf{r}')\mathrm{d}\tau'\right]+\left[\varphi(\mathbf{r}),\int\varphi^{+}(\mathbf{r}')V(\mathbf{r}')\varphi(\mathbf{r}')\mathrm{d}\tau'\right]$$
$$=\int[\varphi(\mathbf{r}),\varphi^{+}(\mathbf{r}')]\left(-\frac{\hbar^2}{2m}\nabla'^2+V(\mathbf{r}')\right)\varphi(\mathbf{r}')\mathrm{d}\tau'=-\frac{\hbar^2}{2m}\nabla^2\varphi(\mathbf{r})+V(\mathbf{r})\varphi(\mathbf{r}).$$

This result represents the equation of Schrödinger. In an analogous manner, we obtain for the case of variable $\varphi^{+}(\mathbf{r})$:

$$-\mathrm{i}\hbar\dot{\varphi}^{+}=-\frac{\hbar^2}{2m}\nabla^2\varphi^{+}(\mathbf{r})+V(\mathbf{r})\varphi^{+}(\mathbf{r}).$$

One must bear in mind that φ and φ^{+} are operators already.

If there exists some complete orthonormal system of wave functions $\psi_i(\mathbf{r})$, one might determine the field operators at an arbitrary moment in time t through these expansions

$$\varphi(\mathbf{r},t)=\sum_i a_i(t)\psi_i(\mathbf{r}) \text{ and } \varphi^{+}(\mathbf{r},t)=\sum_i \psi_i^{*}(\mathbf{r})a_i^{+}(t),$$

in which a_i and a_i^{+} are the familiar operators of destruction and creation. Through the orthonormality of functions ψ_i, we have

$$a_i(t)=\int\varphi(\mathbf{r},t)\psi_i^{*}(\mathbf{r})\mathrm{d}\tau \text{ and } a_i^{+}(t)=\int\psi_i(\mathbf{r})\varphi^{+}(\mathbf{r},t)\mathrm{d}\tau;$$

consequently,

$$[a_i,a_i^{+}]=\int\psi_i^{*}\psi_i'[\varphi(\mathbf{r},t),\varphi^{+}(\mathbf{r}',t)]\mathrm{d}\tau\,\mathrm{d}\tau'=\int|\psi_i|^2\,\mathrm{d}\tau=1,$$

$$[a_i,a_j]=0 \text{ and } [a_i^{+},a_j^{+}]=0;$$

as $[a_i,a_j^{+}]=0$ at $i\neq j$,

$$[a_i,a_j^{+}]=\delta_{ij}.$$

Quantities a_i and a_i^{+} operate in an abstract space of occupation numbers n_i,

$$a_i|n_i\rangle=\sqrt{n_i}|n_i-1\rangle \text{ and } a_i^{+}|n_i\rangle=\sqrt{n_i+1}|n_i+1\rangle,$$

in which $|n_i\rangle$ are the eigenvectors of quantity $N_i = a_i^+ a_i$;

$$N_i|n_i\rangle = n_i|n_i\rangle.$$

The operator N for total particle number is defined through this expression,

$$N = \int \varphi^+ \varphi \, \mathrm{d}\tau = \sum_{i,j} a_i^+ a_j \int \psi_i^* \psi_j \, \mathrm{d}\tau = \sum_i a_i^+ a_i = \sum_i N_i.$$

For the vacuum state, in which there is no particle, for all values of i this identity is satisfied,

$$N_i|0\rangle = 0.$$

On acting on the vector of the vacuum state with creation operators on a sufficient number of successive occasions, one might obtain the state with an arbitrary number of particles. In a general case,

$$|n_1, n_2, \ldots\rangle = \left\{ \prod_i \frac{1}{\sqrt{n_i!}} (a_i^+)^{n_i} \right\} |0\rangle$$

is the vector describing n_1 particles in state 1, n_2 particles in state 2 and so on. Note that operator a_i^+ creates a particle in a state with wave function $\psi_i(\mathbf{r})$, which incarnates the first quantization. The transfer of the occupation numbers into Hilbert space corresponds to the second quantization.

If $\psi_i(\mathbf{r})$ is the eigenfunction belonging to eigenvalue ε_i of this chosen Hamiltonian,

$$H = \sum_{i,j} a_i^+ a_j \int \psi_i^* \left(-\frac{\hbar^2}{2m} \nabla^2 + V \right) \psi_j \, \mathrm{d}\tau \to \sum_i n_i \varepsilon_i.$$

There is no difficulty in understanding the latter result: it is the energy representation for the field of bosons; n_1 particles occupy the state with energy ε_1, n_2 particles occupy the state with energy ε_2 and so on. All quantities N_i, together with the total number of particles, N, in this case, become the constants of the motion.

In terms of the field operators, one might define the quantity total momentum of the field as

$$\mathbf{P} = \int \varphi^+ \mathbf{p} \varphi \, \mathrm{d}\tau = \sum_{i,j} a_i^+ a_j \int \psi_i^* \mathbf{p} \psi_j \, \mathrm{d}\tau.$$

Let functions ψ_i of the first quantization be the eigenfunctions of the momentum operator $\mathbf{p}$, we have

$$\mathbf{P} \rightarrow \sum_i n_i \mathbf{p}_i,$$

in which $\mathbf{p}_i$ are the eigenvalues of momentum corresponding to functions ψ_i. This heuristic expression in the representation of the occupation numbers indicates that n_1 particles possess momentum $\mathbf{p}_1$, n_2 particles possess momentum $\mathbf{p}_2$ and so on. For free particles, $V = 0$ and, in this case, eigenfunctions ψ_i are the same for both Hamiltonian and momentum.

These aspects of second quantization for a boson field are principal. If Bose-particles of various types figure in a problem, for each type one should introduce its own operators of creation and destruction. The operators belonging to various boson fields commute with each other. Note that all drawn conclusions correspond to one and the same moment in time t. To evaluate the temporal variation of field quantities, one must apply Heisenberg's equations of motion.

We proceed to a fermion field. As for the case of bosons, we introduce field operators in a form of these expansions:

$$\varphi(\mathbf{r}, t) = \sum_i c_i(t)\psi_i(\mathbf{r}) \text{ and } \varphi^+(\mathbf{r}, t) = \sum_i \psi_i^*(\mathbf{r})c_i^+(t).$$

Quantities c_i and c_i^+ are operators, and $\psi_1(\mathbf{r})$, $\psi_2(\mathbf{r})$,... represent orthonormal wave functions in a complete set. For fermions, as is known, one must satisfy Pauli's exclusion principle; we therefore refrain from applying the typical commutation relations but, as Jordan and Wigner showed, we might apply these anticommutation relations:

$$[\varphi(\mathbf{r}, t), \varphi^+(\mathbf{r}', t)]_+ = \varphi(\mathbf{r}, t)\varphi^+(\mathbf{r}', t) + \varphi^+(\mathbf{r}', t)\varphi(\mathbf{r}, t) = \delta(\mathbf{r} - \mathbf{r}'),$$

$$[\varphi(\mathbf{r}, t), \varphi(\mathbf{r}', t)]_+ = 0 \text{ and } [\varphi^+(\mathbf{r}, t), \varphi^+(\mathbf{r}', t)]_+ = 0,$$

$$[c_i, c_j^+]_+ = \delta_{ij}, \quad [c_i, c_j]_+ = [c_i^+, c_j^+]_+ = 0.$$

To elucidate the meaning of the latter expressions, we introduce two possible state vectors for fermions:

$$|1\rangle = \begin{pmatrix} 1 \\ 0 \end{pmatrix} \text{ and } |0\rangle = \begin{pmatrix} 0 \\ 1 \end{pmatrix}.$$

In the former case, a fermion occupies a state and, in the latter, a state is unoccupied. Operators c and c^+, in this case, are expressible through the 2×2 Jordan–Wigner matrices:

$$c = \begin{pmatrix} 0 & 0 \\ 1 & 0 \end{pmatrix} = \frac{1}{2}(\sigma_x - \mathrm{i}\sigma_y) \text{ and } c^+ = \begin{pmatrix} 0 & 1 \\ 0 & 0 \end{pmatrix} = \frac{1}{2}(\sigma_x + \mathrm{i}\sigma_y),$$

in which σ_x and σ_y are Pauli matrices (see section 'Spin and Magnetic Moment'). One sees that

$$cc^+ + c^+c = \begin{pmatrix} 1 & 0 \\ 0 & 1 \end{pmatrix}, \quad [c, c]_+ = [c^+, c^+]_+ = \begin{pmatrix} 0 & 0 \\ 0 & 0 \end{pmatrix}.$$

In a manner analogous to that of the field of bosons, we define the particle-number operator $N = c^+c$:

$$c^+c|1\rangle = 1 \cdot |1\rangle \text{ and } c^+c|0\rangle = 0 \cdot |0\rangle,$$

$$N^2 = c^+cc^+c = c^+(1 - c^+c)c = c^+c = N.$$

Eigenvalues n of the particle-number operator equal 1 and 0, as required through Pauli's principle. Operator c destroys and c^+ creates, a fermion in a given state, i.e.

$$c|1\rangle = |0\rangle \text{ and } c^+|0\rangle = |1\rangle.$$

It is important that

$$c|0\rangle = 0 \text{ and } c^+|1\rangle = 0.$$

Furthermore, we might construct a space of occupation numbers in which the field operators act. For this purpose, one must act with creation operators on the vacuum state. We have

$$|0, 0, \ldots, 1_i, \ldots, 0\rangle = c_i^+|0, 0, \ldots, 0_i, \ldots, 0\rangle,$$

$$|0, 0, \ldots, 1_i, \ldots, 1_j, \ldots, 0\rangle = c_i^+c_j^+|0, 0, \ldots, 0_i, \ldots, 0_j, \ldots, 0\rangle \text{ and so on.}$$

Is the location of various operators c and c^+ before the vacuum-state vector important? The answer is affirmative. To understand this fact, we consider a two-particle state with vector $|1,1\rangle$ and act initially on it with operator c_1. As a result,

$$c_1|1, 1\rangle = c_1c_1^+c_2^+|0, 0\rangle = (1 - c_1^+c_1)c_2^+|0, 0\rangle = |0, 1\rangle.$$

We then act on $|1,1\rangle$ with operator c_2, we have

$$c_2|1, 1\rangle = c_2c_1^+c_2^+|0, 0\rangle = -c_1^+(1 - c_2^+c_2)|0, 0\rangle = -|1, 0\rangle.$$

As we see, through the fact that operators c and c^+ fail to commute (they anticommute) in the latter expression a minus sign appears. Thus, if for a particular order there exists occupied state i, which is located to the left of state j, on the vector on which we act with either operator c_j or operator c_j^+, the minus sign arises; in an opposite case, the plus sign remains. This rule is expressible through simple relations:

$$c_j|n_j\rangle = \vartheta_j n_j|1-n_j\rangle \text{ and } c_j^+|n_j\rangle = \vartheta_j(1-n_j)|1-n_j\rangle,$$

in which

$$\vartheta_j = (-1)^{n_1+n_2+\cdots+n_{j-1}}$$

characterizes the number of occupied states to the left of j. To determine the order of location for various fermion-field operators before the vacuum-state vector, one must write the action of quantities c^+ and c in a form of normal product in which all c^+ appear to the left of c.

In conclusion, we consider a heuristic expression for a Hamiltonian in terms of field Fermi-operators for which we choose the solutions of Schrödinger's one-particle equation, corresponding to eigenvalues ε_i, as wave functions $\psi_i(\mathbf{r})$. We have

$$H = \int \varphi^+\left(-\frac{\hbar^2}{2m}\nabla^2 + V\right)\varphi\,\mathrm{d}\tau = \sum_{i,j} c_i^+ c_j \int \psi_i^*\left(-\frac{\hbar^2}{2m}\nabla^2 + V\right)\psi_j\,\mathrm{d}\tau = \sum_i \varepsilon_i c_i^+ c_i.$$

If $V=0$, then $\varepsilon_i = \mathbf{p}_i^2/2m$; analogously to the conclusions for bosons, the total momentum of the field of fermions is, in this case, given by the formula

$$\mathbf{P} = \sum_i \mathbf{p}_i c_i^+ c_i.$$

Molecules

Considering an atom, we imply a stationary state of all electrons that are in the field of an atomic nucleus. Each electron, in this case, moves in some effective or self-consistent field of the nucleus and the other electrons. This description is, to an extent, approximate. In molecules that include multiple atomic centres, the electrons move in the field of several nuclei: the problem becomes complicated. Some electrons, as before, are mostly connected with a particular nucleus, but other electrons, in a parity manner, become related to various nuclei. The principal supposition that allows one to simplify qualitatively the calculation of the total energy of a molecule is the Born–Oppenheimer approximation. According to this approach, which is sometimes called adiabatic, the motion of the electrons is considered with nuclei fixed in relative locations; through a significant distinction between electronic and nuclear masses, this approximation is practicable when electronic states are well separated in energy. The relative motion of the nuclei determines the molecular rotations and vibrations. For a molecule that comprises N atomic centres, three degrees of freedom pertain to the translation of the molecule as a whole; two or three degrees pertain to rotation of the molecule as a rigid body depending on whether the nuclei in their relative equilibrium locations are collinear or not. For vibrations of non-linear and linear molecules, we have $3N-6$ and $3N-5$ degrees of freedom, respectively.

The translational degrees of freedom are trivially associated with the motion of a molecule taken as a whole, and there is no necessity to devote our attention to it. When discussing a molecule, we imply its intrinsic state, which is characterized by electronic, vibrational and rotational energies. To describe approximately the motion of the electrons, one might apply, for instance, a self-consistent-field method to evaluate an electronic energy. We consider separately the dynamics of nuclear motions. The molecular rotations and vibrations fail to be separable: these motions are mutually related; moreover, they depend on the particular electronic state. Nevertheless, to understand qualitatively the essence of a molecular structure, one might tentatively separate the rotational and vibrational motions. In this case, the molecule bears a resemblance to a typical mechanical top with this rotational Hamiltonian:

$$H_{\mathrm{r}} = \frac{L_A^2}{2I_A} + \frac{L_B^2}{2I_B} + \frac{L_C^2}{2I_C}.$$

The latter expression is directly adopted from the classical mechanics concerning the rotation of a rigid body. Here, $\mathbf{L} = (L_A, L_B, L_C)$ is the angular momentum and I_A, I_B and I_C are the moments of inertia with respect to the principal axes; we number the axes with letters A, B and C. If $I_A \neq I_B \neq I_C \neq I_A$, the molecule is classified as an asymmetric top. If only two moments of inertia are equal, i.e. $I_A = I_B \neq I_C$,

$$H_{\mathrm{r}} = \frac{\mathbf{L}^2}{2I_A} + \frac{1}{2}\left(\frac{1}{I_C} - \frac{1}{I_A}\right)L_C^2$$

and the molecule becomes considered a symmetric top, either prolate if $I_A > I_C$ or oblate if $I_A < I_C$. If all moments of inertia are equal to each other, then

$$H_{\mathrm{r}} = \frac{\mathbf{L}^2}{2I_A}$$

that denotes a spherical top.

One might identify the normal vibrations of the molecule with harmonic oscillators in a set with frequencies ω_i. In this case, a vibrational Hamiltonian has a form

$$H_{\mathrm{os}} = \frac{1}{2}\sum_i p_i^2 + \frac{1}{2}\sum_i \omega_i^2 q_i^2.$$

The first part of this Hamiltonian

$$\sum_i p_i^2/2$$

pertains to the kinetic energy of the nuclei, in which p_i are the momenta conjugate to coordinates q_i. The second part represents the potential energy of nuclear interaction through the electronic field; it equals only approximately

$$\sum_i \omega_i^2 q_i^2/2.$$

Note that normal coordinates q_i are here chosen so that in the Hamiltonian the terms of type $q_i q_j$ with $i \neq j$ disappear; moreover, nuclear masses M_i are absent from this Hamiltonian.

The potential energy of interatomic interaction is generally expressible through an expansion in terms of normal coordinates q_i:

$$V(q) = V_0 + \frac{1}{2}\sum_i \omega_i^2 q_i^2 + \sum_{ijk} a_{ijk} q_i q_j q_k + \sum_{ijk\ell} A_{ijk\ell} q_i q_j q_k q_\ell + \cdots,$$

in which the linear terms disappear through the fact that the first derivative V' equals zero at the equilibrium condition. We see that a set of harmonic oscillators corresponds to a first approximation, which allows one to take into account the interaction of atomic centres with each other. This model is only qualitatively correct: molecular vibrations are anharmonic – they fail to conform to a harmonic law. To describe correctly the vibrations, apart from the quadratic part of the potential energy, one must therefore take into account the normal coordinates to greater than quadratic powers in an expansion of V. These terms additional to H_{os} are defined with anharmonicity coefficients

$$a_{ijk}, A_{ijk\ell}, \ldots$$

and characterize the interactions among various vibrational modes. The calculation of the corresponding corrections is generally performed with a perturbation theory to yield a satisfactory agreement with experiment.

If we interpret quantity V_0, which is independent of coordinates, as the purely electronic energy E_{e} or H_{e}, the Hamiltonian of the molecule assumes a form

$$H = H_{\text{e}} + H_{\text{os}} + H_{\text{r}}.$$

Let us evaluate the order of each quantity appearing here. We begin with the electronic energy. One might estimate it with the aid of Heisenberg's principle of indeterminacy. We have

$$ap \sim \hbar,$$

in which a is a typical linear dimension of a molecule and p is the momentum of the electron; quantity a amounts to a few Å. As a result,

$$E_{\text{e}} \sim \frac{p^2}{m} \sim \frac{\hbar^2}{ma^2} \sim 1 \text{ eV},$$

in which m is the electronic mass.

For the energy of vibrations according to Hamiltonian H_{os}, we have

$$E_{os} = \sum_i \hbar\omega_i \left(n_i + \frac{1}{2} \right), \quad n_i = 0, 1, 2, \ldots,$$

in which n_i are vibrational quantum numbers. We see that $E_{os} \sim \hbar\omega$; ω is the frequency of harmonic vibrations of an oscillator with the mass equal to the average nuclear mass M. The potential energy $M\omega^2 x^2/2$, in which x is the shift of an oscillator from equilibrium, becomes of order E_e at $x = a$; in this case, the molecule becomes dissociated into atoms. Hence,

$$E_{os} \sim \hbar \sqrt{\frac{E_e}{Ma^2}} \sim \sqrt{\frac{m}{M}} E_e;$$

as a typical value of m/M amounts to 10^{-3}–10^{-4},

$$E_{os} \sim 0.1 \text{ eV}.$$

To estimate E_r, we note that $L \sim \hbar$ and $I \sim Ma^2$; consequently,

$$E_r \sim \frac{\hbar^2}{Ma^2} \sim \frac{m}{M} E_e \sim 0.001 \text{ eV}.$$

Quantity E_e is ~ 1000 times E_r and ~ 10 times E_{os}. Such a relation between quantities E_e, E_{os} and E_r of energy is sufficient to enable an approximately separate consideration of the electronic, vibrational and rotational states. The transitions among vibrational and rotational states lie in the infrared region of the electromagnetic spectrum, whereas the frequencies of electronic transitions typically manifest themselves in the visible and ultraviolet regions of the spectrum.

As an example, we consider a diatomic molecule. Such a molecule generally has six degrees of freedom – three translational, two rotational and one associated with the vibration of its atomic centres. According to classical mechanics, the two-body problem – in our case the problem of the interaction between the atomic centres – becomes reducible to the problem of a single body with a reduced mass and the potential energy equal to the electronic energy of interaction of the bodies. We apply this device. Let $\mathbf{r}_1$ and $\mathbf{r}_2$ be the radius vectors of atoms, and M_1 and M_2 be their masses accordingly. Placing the origin at the centre of molecular mass, i.e. assuming

$$\mathbf{r}_1 M_1 + \mathbf{r}_2 M_2 = 0,$$

we exclude from consideration the translational motion of the molecule taken as a whole. We characterize a rotational motion with the aid of the eigenfunctions of squared angular momentum, i.e. spherical harmonics $Y_{\ell k}(\theta,\phi)$, in which θ and ϕ are the spherical angles, and ℓ and k are the corresponding quantum numbers. The corresponding energy is, independent of k,

$$E_r = \frac{\hbar^2 \ell(\ell+1)}{2Mr^2}, \quad \ell = 0, 1, 2, \ldots.$$

Here, $M = M_1 M_2/(M_1 + M_2)$ is the reduced mass of the molecule, $\mathbf{r} = \mathbf{r}_1 - \mathbf{r}_2$ is the radius vector connecting the centres of atoms; Mr^2 is the moment of inertia, $\hbar^2 \ell(\ell+1)$ are the eigenvalues of $\mathbf{L}^2$ and number ℓ is called a rotational quantum number. We approximate an energy of interatomic interaction $V(r)$ in a form of an expansion truncated at second order in terms of displacement $x = r - r_0$:

$$V(r) \approx V(r_0) + \frac{M\omega^2}{2} x^2,$$

in which r_0 is the equilibrium separation corresponding to the equation

$$V'(r_0) = 0.$$

According to this approximation, the vibrational motion of the molecule is described with vector $|n\rangle$ of a harmonic oscillator with mass M and with energy

$$\hbar\omega(n + 1/2),$$

in which ω is the frequency of the vibration and n is a vibrational quantum number running over all positive integers including zero.

Supposing that the considered motions are independent, we obtain for the total energy of the diatomic molecule in state

$$|n\ell k\rangle = Y_{\ell k}(\theta, \phi)|n\rangle$$

this expression,

$$E = E_e + \hbar\omega\left(n + \frac{1}{2}\right) + \frac{\hbar^2 \ell(\ell+1)}{2Mr^2}.$$

With the aid of this formula, one might readily understand the features of the distribution of spectral lines of the molecule. So, taking into account the difference between quantities E_e and E_{os}, for each electronic level E_e, we have vibrational levels in a set with $n = 0, 1, 2, \ldots$ As $E_r \ll E_{os}$, above every vibrational level there are rotational levels. When the electronic state remains invariant, the spectrum of the molecule thus comprises vibration-rotational bands. Moreover, there exists a set of purely electronic terms E_e, E_e', …, which represent the energy levels of an optical electron of the molecule. It is important to note that, in a transition from one electronic state to another, the frequency of the harmonic vibrations and the moment of inertia of the molecule become altered; each electronic state has therefore its own vibration-rotational structure. For the state of a diatomic molecule, the wave function represents a product of the wave function of a purely electronic state and a vibration-rotational vector $|n\ell k\rangle$.

Born–Oppenheimer Approximation

The factorization of a wave function of a molecule into electronic and vibration-rotational parts constitutes the mathematical expression of the Born–Oppenheimer approximation; we consider it in detail. Let the molecule comprises N nuclei with masses M_i and n electrons each with mass m, and the radius vectors of the nuclei and electrons are denoted with $\mathbf{R}_i$ and $\mathbf{r}_j$, respectively. We write the stationary equation of Schrödinger

$$\left(-\frac{\hbar^2}{2}\sum_{i=1}^{N}\frac{\nabla_i^2}{M_i}-\frac{\hbar^2}{2m}\sum_{j=1}^{n}\nabla_j^2+V(\mathbf{R}_1,\mathbf{R}_2,\ldots,\mathbf{R}_N;\mathbf{r}_1,\mathbf{r}_2,\ldots,\mathbf{r}_n)\right)|\Phi\rangle=E|\Phi\rangle$$

for states Φ and energy E of the molecule; V is the total potential energy of electrostatic interaction between each two particles among all nuclei and electrons. Vector $|\Phi\rangle$ is expressible as a product of a nuclear function

$$\psi(\mathbf{R}_1,\mathbf{R}_2,\ldots,\mathbf{R}_N)$$

and an electronic function

$$\varphi_R(\mathbf{r}_1,\mathbf{r}_2,\ldots,\mathbf{r}_n).$$

To define that electronic function, we apply that m/M_i is a small quantity such that one might first neglect the kinetic energy of the nuclei. We have

$$\left(-\frac{\hbar^2}{2m}\sum_{j=1}^{n}\nabla_j^2+V\right)\varphi_R=F_R\varphi_R,$$

in which φ_R, F_R and V depend on nuclear coordinates that become fixed according to supposition.

We substitute $|\Phi\rangle$ having a form $\varphi_R\psi$ into Schrödinger's equation and take into account the purely electronic equation yielding F_R. As a result,

$$-\frac{\hbar^2}{2}\sum_{i=1}^{N}\frac{1}{M_i}\nabla_i^2(\varphi_R\psi)+F_R\varphi_R\psi=E\varphi_R\psi.$$

However,

$$\nabla_i^2(\varphi_R\psi) = \varphi_R\nabla_i^2\psi + \psi\nabla_i^2\varphi_R + 2(\nabla_i\varphi_R)(\nabla_i\psi);$$

therefore,

$$\varphi_R\left(-\frac{\hbar^2}{2}\sum_{i=1}^{N}\frac{1}{M_i}\nabla_i^2 + F_R\right)\psi - \frac{\hbar^2}{2}\sum_{i=1}^{N}\frac{1}{M_i}(\psi\nabla_i^2\varphi_R + 2(\nabla_i\varphi_R)(\nabla_i\psi)) = \varphi_R E\psi.$$

Assuming φ_R to be a real-valued function, we multiply this equation by φ_R and integrate over all electronic variables. Note that

$$\int \varphi_R^2\, \mathrm{d}\tau = 1,$$

$$\int \varphi_R\nabla_i^2\varphi_R\, \mathrm{d}\tau = \frac{1}{2}\nabla_i^2\int \varphi_R^2\, \mathrm{d}\tau - \int (\nabla_i\varphi_R)^2\, \mathrm{d}\tau = -\int (\nabla_i\varphi_R)^2\, \mathrm{d}\tau$$

and

$$\int \varphi_R\nabla_i\varphi_R\, \mathrm{d}\tau = \frac{1}{2}\nabla_i\int \varphi_R^2\, \mathrm{d}\tau = 0,$$

in which $\mathrm{d}\tau = \mathrm{d}\tau_1\mathrm{d}\tau_2\cdots\mathrm{d}\tau_n$ is an element of volume for n electrons. Thus, eventually

$$\left(-\frac{\hbar^2}{2}\sum_{i=1}^{N}\frac{1}{M_i}\nabla_i^2 + F_R + W_R\right)\psi = E\psi,$$

in which

$$W_R = \frac{\hbar^2}{2}\sum_{i=1}^{N}\frac{1}{M_i}\int (\nabla_i\varphi_R)^2\, \mathrm{d}\tau$$

yields a small supplement to F_R. Quantity F_R represents the potential energy of interaction between atomic centres; we treated explicitly this energy when we considered the energy of molecular vibrations. One sees that this approach is mathematically inferred in the factorization of the total wave function of a molecule into electronic and nuclear parts, with a condition that the nuclear equation pertains to a particular electronic state. It becomes clear why, in a transition from one electronic state to another, the frequencies of vibrations and the moment of inertia of the molecule become altered. The physical correctness of this approximation relies on the smallness of amplitudes of the nuclear vibrations relative to equilibrium internuclear separations. For small values of vibrational and rotational quantum numbers such that no vibrational and rotational mode becomes too highly excited, the Born–Oppenheimer approximation hence is valid.

Chemical Bond

Let us discuss the mechanisms of formation of molecules – the types of chemical bond. A chemical bond is generally distinguished into polar and covalent. Of course, in nature, there arise bonds of intermediate types, and also chemical forces having a nature of a weaker interatomic interaction – for instance, van der Waals forces. The polar bond has an electrostatic origin, which might be explained according to a classical point of view. In this case, some atomic centres become a moiety with a positive charge and another moiety with a negative charge. Between these moieties, for which the total or net charge equals zero, the common coulombic attraction arises. Such a bond becomes possible only with the condition that the energy of ionic moieties is less than the energy of neutral atoms; this bond must be energetically beneficial. As an example, one might take a molecule of sodium chloride. Atom Na that has a small first ionization energy releases an electron for the sake of Cl that has a significant electron affinity. As a result, one obtains molecule Na^+Cl^- with a polar bond but, as the electron affinity of Cl is less than the ionization energy of Na, the transfer of charge is incomplete; the resulting bond has about three quarters of the electric dipolar moment that would result from a complete transfer.

The bond of covalent type likewise has an electrostatic origin, for which a classical treatment fails. In this case, the atomic moieties forming a molecule themselves represent neutral systems. Examples are found in the molecules dioxygen O_2 and dihydrogen H_2, which contain atomic centres with identical nuclei. To elucidate the essence of a covalent bond, we consider a molecule of dihydrogen in detail. The vibration-rotational problem, which is here trivial, is not of interest here; the exceptional electronic equation in an adiabatic approximation instead attracts our attention.

Let m and e be the mass and absolute value of charge of an electron. We write a Hamiltonian of the molecule

$$H = H_1 + H_2 + V.$$

Here,

$$H_1 = -\frac{\hbar^2}{2m}\nabla_1^2 - \frac{e^2}{r_{1a}} \quad \text{and} \quad H_2 = -\frac{\hbar^2}{2m}\nabla_2^2 - \frac{e^2}{r_{2b}}$$

are the Hamiltonians of two isolated hydrogen atoms;

$$V = e^2\left(-\frac{1}{r_{1b}} - \frac{1}{r_{2a}} + \frac{1}{r_{12}} + \frac{1}{R}\right)$$

is the electrostatic potential of interparticle interaction that might be considered a perturbation;

$$r_{1a},\ r_{1b},\ r_{2a},\ r_{2b},\ r_{12}\ \text{and}\ R$$

are the corresponding separations between electron 1 and nucleus a, electron 1 and nucleus b, electron 2 and nucleus a, electron 2 and nucleus b, between the two electrons and between the two nuclei. Hamiltonian

$$H^0 = H_1 + H_2$$

yields a zero-order approximation when atoms are separate from each other. As the two atoms approach each other, the influence of interaction V increases. We evaluate it within a framework of the first-order perturbation theory.

At $V = 0$ the energy of a system equals a sum of atomic energies,

$$E_a^0 + E_b^0,$$

and a wave function equals a product of the functions of separate atoms $\varphi_a(\mathbf{r}_{1a})$ and $\varphi_b(\mathbf{r}_{2b})$, which follows from this equation for eigenvalues,

$$H^0(\varphi_a(\mathbf{r}_{1a})\varphi_b(\mathbf{r}_{2b})) = (E_a^0 + E_b^0)\varphi_a(\mathbf{r}_{1a})\varphi_b(\mathbf{r}_{2b}).$$

As both hydrogen atomic centres are indistinguishable in the molecule, then

$$E_a^0 + E_b^0 = 2E^0,$$

in which we omit the index labelling atoms. Interchanging the electrons, we obtain that wave function

$$\varphi_a(\mathbf{r}_{2a})\varphi_b(\mathbf{r}_{1b})$$

also corresponds to energy $2E^0$ and describes the unperturbed state of the molecule. The correct functions of zero-order approximation represent superpositions of

$$\varphi_a(\mathbf{r}_{1a})\varphi_b(\mathbf{r}_{2b})\ \text{and}\ \varphi_a(\mathbf{r}_{2a})\varphi_b(\mathbf{r}_{1b}).$$

We choose them in a form of symmetric and antisymmetric combinations, respectively,

$$\psi_s = C_s(\varphi_a(\mathbf{r}_{1a})\varphi_b(\mathbf{r}_{2b}) + \varphi_a(\mathbf{r}_{2a})\varphi_b(\mathbf{r}_{1b}))$$

and

$$\psi_{as} = C_{as}(\varphi_a(\mathbf{r}_{1a})\varphi_b(\mathbf{r}_{2b}) - \varphi_a(\mathbf{r}_{2a})\varphi_b(\mathbf{r}_{1b})).$$

These real-valued functions are mutually orthogonal. Coefficients C_s and C_{as} are chosen from normalization conditions $\langle\psi_s|\psi_s\rangle = 1$ and $\langle\psi_{as}|\psi_{as}\rangle = 1$. Hence,

$$C_s = 1/\sqrt{2(1+\beta^2)} \text{ and } C_{as} = 1/\sqrt{2(1-\beta^2)},$$

in which

$$\beta = \int \varphi_a(\mathbf{r}_{1a})\varphi_b(\mathbf{r}_{1b})\mathrm{d}\tau_1 = \int \varphi_a(\mathbf{r}_{2a})\varphi_b(\mathbf{r}_{2b})\mathrm{d}\tau_2$$

is the integral of the overlap of wave functions of atoms a and b; $\mathrm{d}\tau_1$ and $\mathrm{d}\tau_2$ are the elements of volume for the first and second electrons.

The first correction to the energy within perturbation theory is defined through a diagonal matrix element of quantity V. For the two unperturbed states, we obtain two values for ΔE,

$$\Delta E_s = \langle\psi_s|V|\psi_s\rangle = \frac{A+B}{1+\beta^2} \text{ and } \Delta E_{as} = \langle\psi_{as}|V|\psi_{as}\rangle = \frac{A-B}{1-\beta^2}.$$

Here, A and B, which depend on R, equal, respectively,

$$A = \iint \varphi_a^2(\mathbf{r}_{1a})\varphi_b^2(\mathbf{r}_{2b})V\,\mathrm{d}\tau_1\,\mathrm{d}\tau_2$$

and

$$B = \iint \varphi_a(\mathbf{r}_{1a})\varphi_b(\mathbf{r}_{1b})\varphi_a(\mathbf{r}_{2a})\varphi_b(\mathbf{r}_{2b})V\,\mathrm{d}\tau_1\,\mathrm{d}\tau_2.$$

Quantity A represents the electrostatic energy of interaction averaged over the states of free atoms. Quantity B yields the value for an exchange energy, the origin of which is a purely quantum-mechanical effect. As

$$B < 0,$$

function $\Delta E_s(R)$ lies lower than function $\Delta E_{as}(R)$. Moreover, the effective potential energy of interaction between atoms, F_R, which is equal to

$$2E^0 + \Delta E(R),$$

has a minimum for the case of only symmetric wave function ψ_s. Therefore, ψ_s leads to formation of a molecule, whereas in state ψ_{as} atoms repel each other.

There remains a question about the spin state of the molecule of dihydrogen. To reply to this query, we apply Pauli's exclusion principle. The total electronic wave function represents a product of space function $\psi_s(1,2)$ and spin function $\chi(\mathbf{s}_1,\mathbf{s}_2)$, in which $\mathbf{s}_1$ and $\mathbf{s}_2$ are the spin angular momenta of the first and second electrons, respectively. This function must be antisymmetric, i.e.

$$\psi_s(1,2)\chi(\mathbf{s}_1,\mathbf{s}_2) = -\psi_s(2,1)\chi(\mathbf{s}_2,\mathbf{s}_1),$$

in which arguments 1 and 2 between parentheses following ψ_s designate the spatial coordinates of electrons. As $\psi_s(1,2) = \psi_s(2,1)$, then

$$\chi(\mathbf{s}_1,\mathbf{s}_2) = -\chi(\mathbf{s}_2,\mathbf{s}_1).$$

The spin function is thus antisymmetric; we represent it in a form

$$\chi(\mathbf{s}_1,\mathbf{s}_2) = \frac{1}{\sqrt{2}}(\chi_1(\mathbf{s}_1)\chi_2(\mathbf{s}_2) - \chi_1(\mathbf{s}_2)\chi_2(\mathbf{s}_1)),$$

in which χ_1 and χ_2 are the orthonormal one-particle spin functions. In the molecule of dihydrogen, the chemical bond becomes possible in the case that the spins of electrons are only oppositely directed. Such a situation is characteristic of any covalent bond. One might consider that this bond arises through the electrons with antiparallel spins. Note that the mutual orientation of nuclear spins might be arbitrary. If the spins of nuclei are parallel to each other, one has an ortho state of the molecule of dihydrogen; in the opposite case, when the spins are oppositely directed, we treat a para state.

Questions of Symmetry

The symmetry of a geometric body implies the presence of a compatibility of that body for transformations according to rotation by an angle about an axis and reflection, like that of a mirror, at a plane. For each body, the total set of symmetry elements, including an identity transformation, forms a group. The successive application of multiple transformations is expressed through the product of the corresponding elements of the group; the product becomes an element of the same group and fails generally to possess the property of commutativity. A symmetry element resembles some operation to which in a given linear space one might ascribe an operator or matrix. The operators of all elements of the group in a set constitute a group representation with a dimension that equals the dimension of a given space. By definition, the order of the group is equal to the total number of its elements. Let us consider the principal symmetry operations.

The operation rotation about an axis by angle $2\pi/n$, in which $n = 1, 2, 3,\ldots$, is commonly designated by symbol C_n. The body possesses an n-fold axis of symmetry if transformation C_n transfers the body to the state that is identical with the original state. Performing successively two operations C_n, we obtain transformation

C_nC_n or C_n^2 corresponding to the rotation by angle $2(2\pi/n)$. Transformations $C_n^3, C_n^4, \ldots$ and C_n^{n-1} are thus rotations by angles

$$3 \cdot \frac{2\pi}{n},\ 4 \cdot \frac{2\pi}{n}, \ldots \text{ and } (n-1) \cdot \frac{2\pi}{n}.$$

Obviously, C_n^n performs the identity transformation; the latter is generally denoted I.

The transformation reflection at a plane, at which the body goes over into itself, is designated with symbol σ; in this case, the plane is called the plane of symmetry. Applying this operation twice, we obtain that

$$\sigma^2 = I.$$

If some transformation simultaneously consists of reflection σ and rotation C_n, one should distinguish between reflection σ_v at a plane containing this axis and reflection σ_h at a plane perpendicular to the axis of rotation. We represent these operations in forms

$$\sigma_v C_n \text{ and } \sigma_h C_n,$$

respectively.

The case of successive application of σ_h and C_n constitutes special interest, as it leads to the concept of a rotation–reflection transformation S_n. We have by definition

$$S_n = \sigma_h C_n = C_n \sigma_h.$$

In particular, S_2 is the transformation constituting inversion i,

$$i = \sigma_h C_2 = S_2;$$

this operation allows one to convert vector $\mathbf{r}$ of each point of a body into vector $-\mathbf{r}$. The body possesses an n-fold rotation–reflection axis of symmetry if rotation by angle $2\pi/n$ about a given axis and subsequent reflection at a plane perpendicular to this axis transform the body into a state indistinguishable from the original one; if $n = 2$, the body possesses a centre of symmetry.

In quantum mechanics, considering a system of finite size – for instance, a molecule or its moiety – each symmetry operation might be determined through a corresponding coordinate transformation that leaves invariant a Hamiltonian of the system. In this case, it is convenient to represent the symmetry elements of the system in a total set by a *symmetry group*. The most important groups in physics are the so-called point groups. Any transformation belonging to such a group retains unaltered at least one point of the system in a space; all axes and planes of symmetry of the system must thus have at least one common point of intersection.

For example, the point groups of symmetry operations are effectively applicable to classify the normal vibrations and electronic states of some molecules.

Point Groups

Let us introduce the concept of a class of *conjugate elements* of a group. Two arbitrary elements g_1 and g_2 are called mutually conjugate if

$$g_1 = g_3 g_2 g_3^{-1},$$

in which g_3 belongs to the same group and g_3^{-1} is the element that is reciprocal to g_3; $g_3 g_3^{-1} = I$. If

$$g_1 = g_3 g_2 g_3^{-1} \text{ and } g_2 = g_5 g_4 g_5^{-1},$$

then

$$g_1 = (g_3 g_5) g_4 (g_3 g_5)^{-1};$$

as g_4, g_5 and $g_3 g_5$ are also elements of the group,

$$g_1,\ g_2 \text{ and } g_4$$

are conjugate to each other. The mutually conjugate elements of the group in a total set, by definition, constitute the class. Any group might thus be separated into classes, each of which is determined by one of its elements. The order of the class equals the number of its elements. Identity or unity element I invariably occupies a separate class.

The elements of the point groups of symmetry are rotations and reflections. It is intuitively clear that to one class of some group belong only those rotations through the same angle, the axes of which might be transformed into each other through appropriate transformation of this group. This conclusion is extensible to reflections in an analogous manner. If among the elements of the group there exists an operation that transforms one plane of symmetry into another, reflections with respect to these planes then enter one class. We consider, for instance, an element g of a group that effects a rotation about axis 1, and conjugate to it element

$$f = u g u^{-1},$$

in which u is also an element of the group. We show that

$$g \text{ and } f$$

produce rotations through the same angles. Let u acting on axis 1 transform it into axis 2. On acting with element f on axis 2, we have

$$ug(u^{-1} \rightarrow \text{axis } 2) = u(g \rightarrow \text{axis } 1) = (u \rightarrow \text{axis } 1) = \text{axis } 2.$$

We see that axis 2 retains unchanged; consequently, f is the rotation. Furthermore, we suggest that g is a rotation about an n-fold axis, then

$$g = C_n \text{ and } g^n = I.$$

However,

$$f^n = (ugu^{-1})^n = ug^n u^{-1},$$

hence

$$f^n = I.$$

The intuitive suggestion is proved: elements g and f belonging to one class perform rotations through one and the same angle.

Following the notation of Schönflies, we consider the principal point groups.

Groups $\boldsymbol{C}_n$. The point group of type $\boldsymbol{C}_n$ comprises the rotations about an n-fold axis of symmetry. Each of its elements

$$C_n, C_n^2, C_n^3, \ldots, C_n^n$$

forms a separate class. The case $n = 1$ corresponds to an absence of any symmetry: $\boldsymbol{C}_1$ consists of only one element, I. As an instance of group $\boldsymbol{C}_n$, we present the molecule H_2O_2. This molecule in the case of a non-planar configuration possesses a two-fold axis of symmetry (Figure 1.1).

Groups $\boldsymbol{C}_{nh}$. An n-fold axis of symmetry and a plane of symmetry perpendicular to this axis form point group $\boldsymbol{C}_{nh}$. In fact, $\boldsymbol{C}_{nh}$ is a result of a *direct* product of group $\boldsymbol{C}_n$ and group $\boldsymbol{\sigma}_h$ consisting by definition of two elements I and σ_h. The elements of a direct product

$$\boldsymbol{C}_n \times \boldsymbol{\sigma}_h$$

are obtainable through a multiplication of each element from $\boldsymbol{C}_n$ with each element from $\boldsymbol{\sigma}_h$. Thus, $\boldsymbol{C}_{nh}$ contains n rotations,

$$C_n, C_n^2, C_n^3, \ldots, C_n^n$$

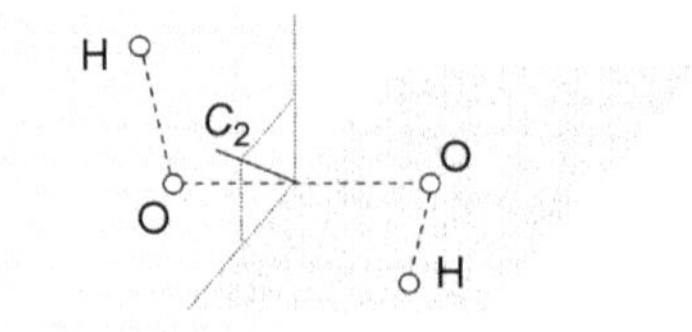

Figure 1.1 Non-planar configuration of H_2O_2 with two-fold axis of symmetry.

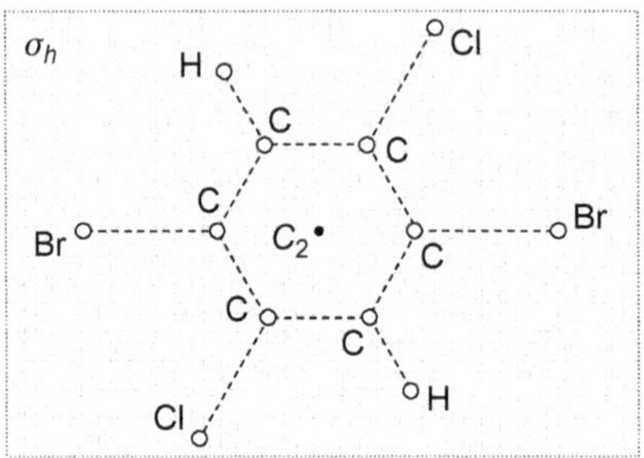

Figure 1.2 $C_6H_2Cl_2Br_2$ molecule representing point group $\boldsymbol{C}_{2h}$.

and n rotary reflections,

$$C_n\sigma_h, C_n^2\sigma_h, C_n^3\sigma_h, \ldots, C_n^n\sigma_h.$$

The molecule $C_6H_2Cl_2Br_2$ (Figure 1.2), for instance, belongs to the group of symmetry $\boldsymbol{C}_{nh}$; here $n = 2$.

Groups $\boldsymbol{C}_{nv}$. An n-fold axis of symmetry and n planes of symmetry through this axis represent point group $\boldsymbol{C}_{nv}$. The elements of $\boldsymbol{C}_{nv}$ are n rotations

$$C_n, C_n^2, C_n^3, \ldots, C_n^n$$

and n *vertical* reflections σ_v. The neighbouring planes intersect each other along the axis of symmetry at angle π/n. In the case of odd values of n, each *vertical* plane might be transformed into any other with the aid of an operation corresponding to rotation C_n through angle $2\pi/n$; all reflections thereby enter one class. For even n, reflections become distributed into two classes, as successive rotations C_n allow one to make compatible with each other only half of all planes.

One sees that the operation of reflection σ_v with subsequent rotation C_n^k through angle $2\pi k/n$ is equivalent to the transformation of rotation C_n^{-k} through angle $2\pi k/n$ in the reverse direction with subsequent reflection σ_v, i.e.

$$C_n^k\sigma_v = \sigma_v C_n^{-k},$$

where from

$$C_n^k = \sigma_v C_n^{-k}\sigma_v^{-1},$$

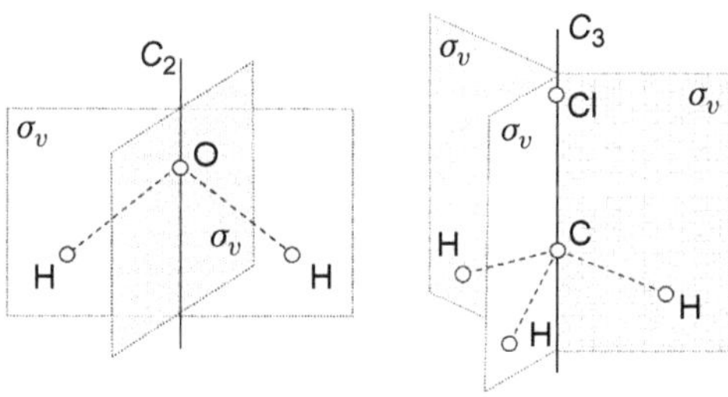

Figure 1.3 Molecules H_2O and CH_3Cl as instances of point group $\boldsymbol{C}_{nv}$.

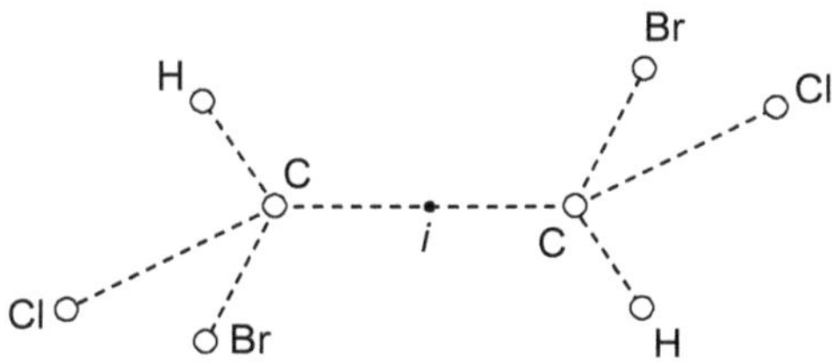

Figure 1.4 Molecule $C_2H_2Br_2Cl_2$ illustrating point group $\boldsymbol{S}_2$.

in which $\sigma_v^{-1} = \sigma_v$. Through the presence of n vertical planes of symmetry, all rotations become thus conjugate in pairs. Noticing that $C_n^{-k} = C_n^{n-k}$, we conclude that elements C_n^k and C_n^{n-k} enter one class. The molecules H_2O and CH_3Cl (Figure 1.3) might serve as examples of point group $\boldsymbol{C}_{nv}$.

Groups $\boldsymbol{S}_{2n}$. The point group of type $\boldsymbol{S}_{2n}$ is formed by rotary reflections

$$S_{2n}, S_{2n}^2, S_{2n}^3, \ldots, S_{2n}^{2n}$$

about a $2n$-fold rotation–reflection axis of symmetry. Group $\boldsymbol{S}_2$ consisting of two elements I and S_2 corresponds to the presence of a centre of symmetry; as $S_2 = i$, this group is commonly denoted with $\boldsymbol{i}$. An example of $\boldsymbol{S}_{2n}$ is the molecule $C_2H_2Br_2Cl_2$ (Figure 1.4). The case $\boldsymbol{S}_{2n+1}$ has no practical interest, as

$$S_{2n+1} = C_{2n+1}\sigma_h, S_{2n+1}^2 = C_{2n+1}^2, \ldots, S_{2n+1}^{2n+1} = \sigma_h,$$

$$S_{2n+1}^{2n+2} = C_{2n+1}, \ldots, S_{2n+1}^{4n+2} = I,$$

such that

$$\boldsymbol{S}_{2n+1} = \boldsymbol{C}_{2n+1} \times \boldsymbol{\sigma}_h.$$

A $2n + 1$-fold rotation–reflection axis thus fails to be an independent symmetry element and is equivalent to a combination of a $2n + 1$-fold axis of symmetry and a plane of symmetry perpendicular to this axis.

Groups $\boldsymbol{D}_n$. An n-fold axis of symmetry and n two-fold axes of symmetry perpendicular to the n-fold axis represent point group $\boldsymbol{D}_n$. Together with the elements of group $\boldsymbol{C}_n$, $\boldsymbol{D}_n$ contains the rotations through angle π with respect to n *horizontal*

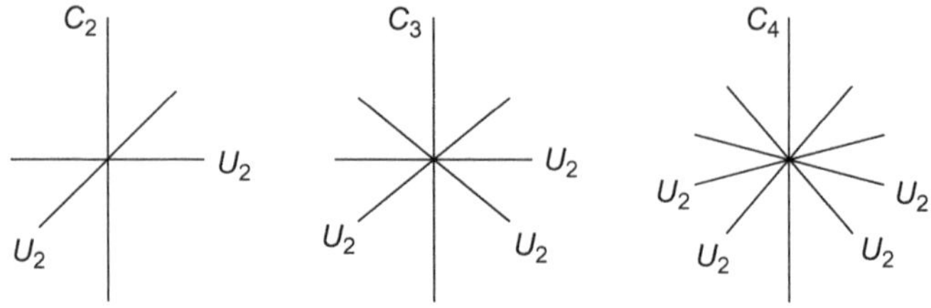

Figure 1.5 Systems of axes of point groups $\boldsymbol{D}_2$, $\boldsymbol{D}_3$ and $\boldsymbol{D}_4$.

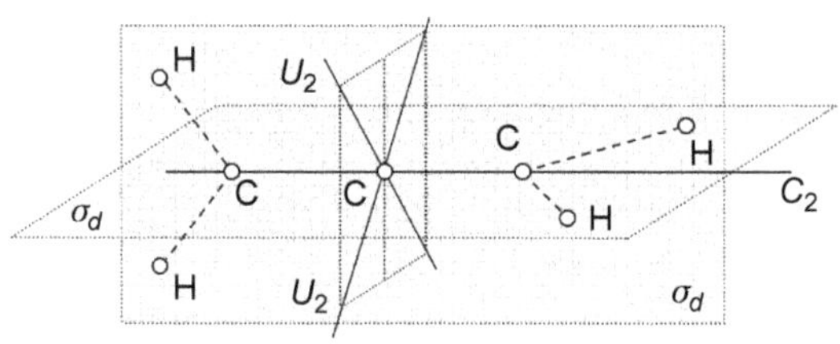

Figure 1.6 Allene molecule as instance of point group $\boldsymbol{D}_{nd}$.

axes (Figure 1.5). The operation of rotation about a *horizontal* axis to differentiate it from operation of rotation C_2 with respect to a *vertical* axis, we designate as U_2. In an abstract meaning, transformation U_2 of group $\boldsymbol{D}_n$ uniquely corresponds to the transformation of reflection σ_v of group $\boldsymbol{C}_{nv}$. On acting on an n-fold axis, U_2 reverses its direction. Rotations C_n^k and C_n^{n-k}, which are in direct and opposite directions respectively, become therefore conjugate to each other; they hence belong to one class. All rotations about the horizontal axes of group $\boldsymbol{D}_n$ become distributed into two classes if n is even, but they enter one class in the case of odd values of n.

Groups $\boldsymbol{D}_{nd}$. The point group of type $\boldsymbol{D}_{nd}$ contains an n-fold axis of symmetry, n two-fold horizontal axes of symmetry and n *diagonal* planes of symmetry. Each *diagonal* plane contains, by definition, a vertical axis and bisects the angle between two neighbouring horizontal axes; the reflection with respect to such a plane is denoted σ_d. The elements of group $\boldsymbol{D}_{nd}$ are thus n rotations

$$C_n, C_n^2, C_n^3, \ldots, C_n^n,$$

n rotations U_2, n reflections σ_d and n operations $U_2\sigma_d$. All reflections σ_d enter one class, as all diagonal planes are combined with each other through transformations U_2. The rotations about n two-fold horizontal axes also belong to one class; all axes are obtainable from one with the aid of reflections σ_d. An example of group $\boldsymbol{D}_{nd}$ is the allene molecule C_3H_4 (Figure 1.6).

Groups $\boldsymbol{D}_{nh}$. An n-fold axis and n vertical planes of the point group of symmetry $\boldsymbol{C}_{nv}$ plus a *horizontal* plane of symmetry perpendicular to the axis form point group $\boldsymbol{D}_{nh}$. The presence of planes of symmetry perpendicular to each other causes the appearance of n two-fold horizontal axes of symmetry, which are directed along

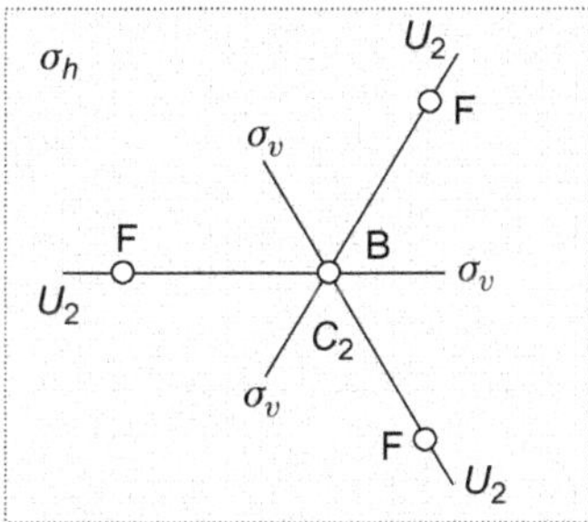

Figure 1.7 Molecule BF_3 representing point group $\boldsymbol{D}_{nh}$.

the lines of intersection of the horizontal and vertical planes. Group $\boldsymbol{D}_{nh}$ thus contains n rotations

$$C_n, C_n^2, C_n^3, \ldots, C_n^n,$$

n rotations U_2, n reflections σ_v and n rotary reflections

$$C_n\sigma_h, C_n^2\sigma_h, C_n^3\sigma_h, \ldots, C_n^n\sigma_h.$$

For even values of n, one might represent the group $\boldsymbol{D}_{nh}$ in a form of a direct product of $\boldsymbol{D}_n$ and $\boldsymbol{i}$, i.e.

$$\boldsymbol{D}_{nh} = \boldsymbol{D}_n \times \boldsymbol{i};$$

if n is odd, then

$$\boldsymbol{D}_{nh} = \boldsymbol{D}_n \times \boldsymbol{\sigma}_h.$$

As an instance of $\boldsymbol{D}_{nh}$, we present the molecule BF_3 (Figure 1.7).

Groups $\boldsymbol{T}$, $\boldsymbol{T}_h$ *and* $\boldsymbol{T}_d$. Point group $\boldsymbol{T}$ is the complex of all rotations about three mutually perpendicular two-fold axes of symmetry and four three-fold axes of symmetry, which stipulate the compatibility of a regular tetrahedron with itself. The two-fold axes bisect the angles that are formed by the three-fold axes of symmetry; they are hence obtainable one from another through operations C_3. Rotations C_2 convert to each other the four three-fold axes. The twelve elements of the group $\boldsymbol{T}$ are thus distributed into these four classes:

{identity element I}, {three rotations C_2}, {four rotations C_3} and {four rotations C_3^2};

the elements belonging to one class are terminated with the braces.

The system of axes of symmetry of a regular tetrahedron and a centre of symmetry in a set form point group $\boldsymbol{T}_h$. By definition,

$$\boldsymbol{T}_h = \boldsymbol{T} \times \boldsymbol{i};$$

twenty-four elements of group $\boldsymbol{T}_h$ are consequently distributed into eight classes:

$$\{I\}, \{3C_2\}, \{4C_3\}, \{4C_3^2\}, \{i\}, \{3C_2 i\}, \{4C_3 i\} \text{ and } \{4C_3^2 i\}.$$

Here,

$$C_2 i = C_2^2 \sigma_h = \sigma_h,$$

$$C_3 i = C_3 C_2 \sigma_h = C_6^5 \sigma_h = C_6^{-1} \sigma_h = S_6^{-1}$$

and

$$C_3^2 i = C_3 C_3 C_2 \sigma_h = C_6 \sigma_h = S_6,$$

a simultaneous application of operations inversion i and rotation C_n^k, in which $k = 1, 2, \ldots, n-1$, yields two rotations C_n^k and C_2 through angles $2\pi k/n$ and π, respectively, about one and the same axis with subsequent reflection σ_h at the plane perpendicular to the axis.

Point group $\boldsymbol{T}_d$ represents the total complex of symmetry elements of the regular tetrahedron. Together with the system of axes of symmetry of group $\boldsymbol{T}$, in $\boldsymbol{T}_d$ there appear six planes of symmetry, each of which is formed by a pair of intersecting three-fold axes; every plane also contains one of the three two-fold axes of symmetry. To show that two-fold axes become four-fold rotation−reflection axes, we let numerals 1, 2 and 3 number the two-fold axes of symmetry;

$$C_2^{(1)}, C_2^{(2)} \text{ and } C_2^{(3)}$$

denote the transformations of rotations about these axes, respectively. We consider element

$$g = C_2^{(2)} \sigma$$

of group $\boldsymbol{T}_d$, in which axis 2 bisects the angle between mutually perpendicular planes of symmetry (Figure 1.8); σ is the reflection at one such plane. One might

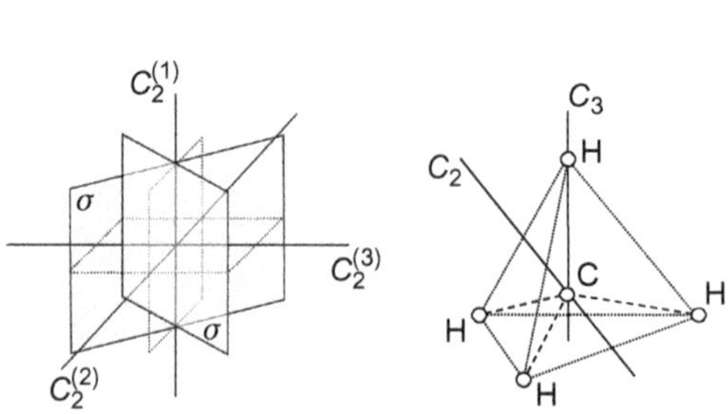

Figure 1.8 Illustrations of point group $\boldsymbol{T}_d$.

represent rotation $C_2^{(2)}$ in a form of a product of two reflections σ_h and σ_v at planes that intersect each other along axis 2 at angle $\pi/2$; we have

$$g = \sigma_h \sigma_v \sigma.$$

Furthermore,

$$\sigma_v \sigma = C_4^{(1)},$$

as the planes of reflections σ_v and σ intersect each other along axis 1 at angle $\pi/4$. As a result,

$$g = \sigma_h C_4^{(1)} = S_4^{(1)},$$

axis 1 becomes a four-fold rotation–reflection axis. For axes 2 and 3, the conclusions apply analogously.

Apart from the elements of group $\boldsymbol{T}$ and six reflections σ, in $\boldsymbol{T}_d$ we thus have six rotary reflections

$$S_4 \text{ and } S_4^{-1}.$$

As

$$\sigma S_4 \sigma^{-1} = \sigma \sigma_h C_4 \sigma^{-1} = \sigma_h \sigma C_4 \sigma^{-1} = \sigma_h C_4^{-1} = S_4^{-1},$$

rotations S_4 and S_4^{-1} are conjugate to each other and enter one class. Analogously, through this relation,

$$\sigma C_3 \sigma^{-1} = C_3^{-1},$$

there arise the mutually conjugate rotations

$$C_3 \text{ and } C_3^{-1}.$$

The planes of symmetry are obtainable from one another through rotations C_3 and C_2, hence they all belong to one class. Accordingly, group $\boldsymbol{T}_d$ of symmetry of the regular tetrahedron has these five classes:

$$\{I\}, \{3C_2\}, \{4C_3 \text{ and } 4C_3^{-1}\}, \{6\sigma\} \text{ and } \{3S_4 \text{ and } 3S_4^{-1}\}.$$

An illustration of point group $\boldsymbol{T}_d$ is the methane molecule CH_4 (see Figure 1.8).

Groups $\boldsymbol{O}$ and $\boldsymbol{O}_h$. Six two-fold axes of symmetry, four three-fold axes of symmetry and three mutually perpendicular four-fold axes of symmetry form point group $\boldsymbol{O}$. All axes of the same type are obtainable from one another and, moreover,

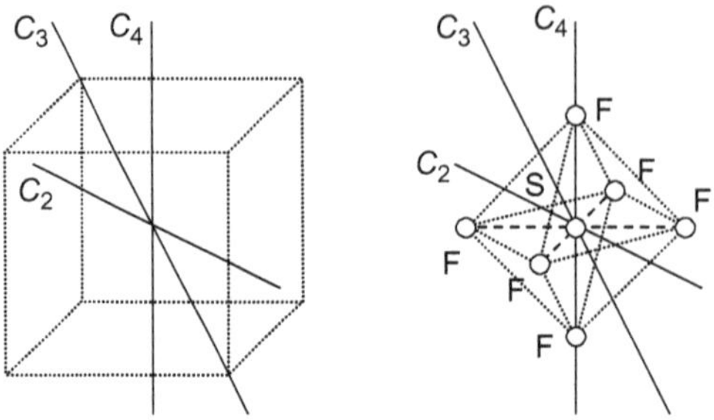

Figure 1.9 Illustrations of point groups $\boldsymbol{O}$ and $\boldsymbol{O}_h$.

might be reoriented in the opposite direction with the aid of appropriate combinations of rotations that belong to the group. The rotations through the same angles about the axes of each type in the direct and opposite directions, consequently, enter one class. Accordingly, group $\boldsymbol{O}$, comprising twenty-four elements, contains these five classes:

$$\{I\}, \{6C_2\}, \{4C_3 \text{ and } 4C_3^{-1}\},\ \{3C_4 \text{ and } 3C_4^{-1}\} \text{ and } \{3C_4^2\}.$$

One sees that the rotations about the axes of symmetry of point group $\boldsymbol{O}$ stipulate the compatibility of a cube with itself (Figure 1.9).

Point group $\boldsymbol{O}_h$ is the complex of all axes of symmetry of octahedral group $\boldsymbol{O}$ and a centre of symmetry. As

$$\boldsymbol{O}_h = \boldsymbol{O} \times \boldsymbol{i},$$

forty-eight elements of the group are distributed into ten classes:

$$\{I\}, \{6C_2\}, \{4C_3 \text{ and } 4C_3^{-1}\}, \{3C_4 \text{ and } 3C_4^{-1}\}, \{3C_4^2\}, \{i\}, \{6C_2 i\},$$
$$\{4C_3 i \text{ and } 4C_3^{-1} i\}, \{3C_4 i \text{ and } 3C_4^{-1} i\} \text{ and } \{3C_4^2 i\};$$

$$C_2 i = C_2^2 \sigma_h = \sigma_h,$$

$$C_3 i = C_3 C_2 \sigma_h = C_6^5 \sigma_h = S_6^5, \quad C_3^{-1} i = C_3^{-1} C_2 \sigma_h = C_6 \sigma_h = S_6,$$

$$C_4 i = C_4 C_2 \sigma_h = C_4^3 \sigma_h = S_4^3, \quad C_4^{-1} i = C_4^{-1} C_2 \sigma_h = C_4 \sigma_h = S_4$$

and

$$C_4^2 i = C_4^2 C_2 \sigma_h = \sigma_h,$$

in which each reflection σ_h arises at a plane perpendicular to the corresponding axis. For instance, the molecule SF_6 (see Figure 1.9) might serve as an illustration of point group $\boldsymbol{O}_h$.

Groups $\boldsymbol{Y}$ *and* $\boldsymbol{Y}_h$. Point group $\boldsymbol{Y}$ is the set of all rotations that stipulate the compatibility of a regular icosahedron with itself. In $\boldsymbol{Y}_h$, in addition to the axes of symmetry $\boldsymbol{Y}$, there appears a centre of symmetry:

$$\boldsymbol{Y}_h = \boldsymbol{Y} \times \boldsymbol{i}.$$

Fullerene C_{60} might serve as an example of group $\boldsymbol{Y}_h$. This molecule represents a closed spheroidal surface, in which the carbon atoms occupy the vertices of twenty regular hexagons and twelve regular pentagons.

Classification of States According to Symmetry

Let operator G corresponds to each element g of the symmetry group of a molecule with Hamiltonian H. For the molecule, we write the stationary Schrödinger equation

$$H\varphi_k = E\varphi_k,$$

in which $\varphi_1, \varphi_2, \ldots, \varphi_r$ are wave functions numbering r in a set, each of which corresponds to a value of energy E. Through the invariance of this equation with respect to all transformations g, we have

$$G(H\varphi_k) = H(G\varphi_k) = E(G\varphi_k).$$

Function $G\varphi_k$, thus, also corresponds to value E and might be represented in a form of linear combination,

$$G\varphi_k = \sum_{\ell=1}^{r} G_{\ell k}\varphi_\ell.$$

The complex of matrix elements $G_{\ell k}$ constitutes a *representation* of transformation g. The dimension of the representation equals number r of linearly independent functions $\varphi_1, \varphi_2, \ldots, \varphi_r$ that form a *basis* of the given representation.

The matrices, in a set, of all elements of the group under consideration constitute the group representation. The basis functions, with the aid of which one constructs the representation, might generally fail to be direct solutions of the Schrödinger equation; the choice of these functions is arbitrary. To various sets of basis functions, hence correspond various representations. If two bases are related to each other through a linear transformation, the representations corresponding to them are mutually equivalent. Operators G and G' of the two mutually equivalent representations are connected through this simple relation

$$G' = TGT^{-1},$$

in which T is a linear operator yielding the correspondence between the two representations for each element g of the group.

A representation of the group might be either reducible or irreducible. One might decompose the basis of the reducible representation of dimension r with the aid of an appropriate linear transformation into the sets of functions

$$\underbrace{\varphi_1, \varphi_2, \ldots, \varphi_{s_1}}_{s_1,} \quad \underbrace{\varphi_{s_1+1}, \varphi_{s_1+2}, \ldots, \varphi_{s_1+s_2}}_{s_2,} \quad \ldots \text{ and } \underbrace{\varphi_{r-s_p+1}, \varphi_{r-s_p+2}, \ldots, \varphi_r}_{s_p,}$$

in which

$$s_1 + s_2 + \cdots + s_p = r;$$

the functions belonging to various sets at all operations G of the given group do not mix with each other. The representation is thus *reducible* if

$$G\varphi_k = \sum_{\ell=1}^{s_1} \langle \varphi_\ell | G | \varphi_k \rangle \varphi_\ell, \quad k = 1, 2, \ldots, s_1;$$

$$G\varphi_i = \sum_{\ell=s_1+1}^{s_1+s_2} \langle \varphi_\ell | G | \varphi_i \rangle \varphi_\ell, \quad i = s_1 + 1, s_1 + 2, \ldots, s_1 + s_2;$$

$$\vdots$$

$$G\varphi_m = \sum_{\ell=r-s_p+1}^{r} \langle \varphi_\ell | G | \varphi_m \rangle \varphi_\ell, \quad m = r - s_p + 1, r - s_p + 2, \ldots, r.$$

If there is no linear transformation according to which the basis becomes decomposed into sets of mutually transforming functions, the representation is *irreducible*.

Each reducible representation contains irreducible parts of determinate number. The decomposition into irreducible parts is performed in accordance with Burnside's elegant formula

$$s_1^2 + s_2^2 + \cdots + s_p^2 = S,$$

in which S is the order of the group, s_1, s_2,..., s_p are the dimensions of all p irreducible non-equivalent representations; number p equals the number of classes in the group. For given values of p and S, the decomposition of Burnside is unique. For instance, if

$$p = 10 \text{ and } S = 48,$$

then

$$s_1 = s_2 = s_3 = s_4 = 1, \ s_5 = s_6 = 2 \ \text{and} \ s_7 = s_8 = s_9 = s_{10} = 3,$$

as

$$1^2 + 1^2 + 1^2 + 1^2 + 2^2 + 2^2 + 3^2 + 3^2 + 3^2 + 3^2 = 48;$$

this case, as one easily sees, corresponds to point group $\boldsymbol{O}_h$, which thus has four one-dimensional, two two-dimensional and four three-dimensional irreducible representations.

In physical applications of the theory of point groups, the irreducible representations corresponding to particular symmetry types have specific designations. Letters A and B designate the one-dimensional representations; they correspond to non-degenerate symmetry types. The basis functions of type A are symmetric with respect to rotations about the principal n-fold axis of symmetry, whereas the functions of type B are antisymmetric with respect to these rotations. The two-dimensional and three-dimensional representations are denoted by letters E and F, respectively; they correspond to doubly and triply degenerate symmetry types. Numerals 1 and 2 as subscripts indicate the symmetry with regard to reflection σ_v at the vertical plane through the principal axis. For example, with respect to operation σ_v, the basis functions of type A_1 are symmetric and functions A_2 are antisymmetric. The prime and double prime denote the symmetry with regard to reflection σ_h at the plane perpendicular to the principal axis. For instance, the basis functions of type A' are symmetric and functions A'' are antisymmetric with respect to operation σ_h. If X is one of representations A, B, E or F, symbols

$$X_g \text{ and } X_u$$

correspond to the representations that are respectively even and odd with respect to the transformation of inversion i. For other symmetry types, there apply analogous designations A_2', B_{1u}, E_g, F_{1u} and so on.

In accordance with the symmetry types of the pertinent point groups, the normal vibrations of molecules become classifiable. One might schematically represent the classification in a following manner. We first reveal all *totally symmetric* vibrations, which maintain the symmetry of the equilibrium configuration of a molecule. Then, discarding one by one the elements of the symmetry group of the molecule, we define other vibrations. The non-degenerate vibrations correspond to the one-dimensional representations; such vibrations are called simple. The normal coordinates of degenerate vibrations are classified through the irreducible representations of dimension r; number r equals the degree of degeneracy of the corresponding vibration. As an example, we consider the ammonia molecule NH_3. Ammonia belongs to point group $\boldsymbol{C}_{3v}$, six elements of which are distributed into three classes

$$\{I\},\ \{C_3 \text{ and } C_3^{-1}\} \text{ and } \{3\sigma_v\}.$$

According to Burnside's relation,

$$1^2 + 1^2 + 2^2 = 6,$$

$\boldsymbol{C}_{3v}$ has two one-dimensional and one two-dimensional representations. The possible symmetry types of normal vibrations of NH_3 are thus A and E. The four frequencies

$$\omega_1, \omega_2, \omega_3 \text{ and } \omega_4$$

correspond classically to two totally symmetric and two doubly degenerate vibrations of the molecule (Figure 1.10); the figure shows only one component of each vibration of E-type.

Like vibrations, the electronic states of molecules are classified according to the irreducible representations of the corresponding point groups. The displacements of nuclei undergoing internal vibrations of a molecule are, as a rule, sufficiently small that the symmetry of the equilibrium configuration of the molecule is approximately maintained, and the classification of the electronic energy levels might be obtained for fixed nuclei. The electronic wave functions corresponding to one and the same value of energy belong to one and the same symmetry type; they are consequently transformed through each other with the same irreducible representation. The degree of degeneracy of the electronic states is equal to the dimension of the representation. For instance, the electronic states of molecule NH_3, which possesses the symmetry of point group $\boldsymbol{C}_{3v}$, might be both non-degenerate and doubly degenerate.

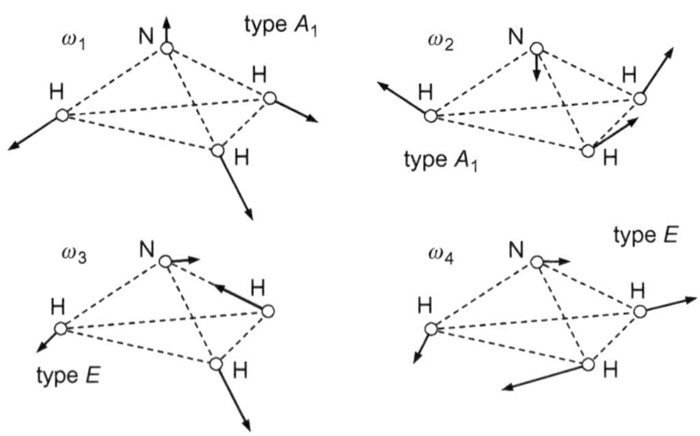

Figure 1.10 Normal vibrations of NH_3.

2 The Evolution of Perturbation Theory

Preamble

Vibrational phenomena have always fascinated scientists and engineers. A molecule constitutes a vibrational system of an important class that is our main concern here. High-resolution infrared absorption spectra provide information about the distribution of vibration–rotational energy levels and the transition probabilities of real molecules. Spectral lines command physical interest through their interpretation with the aid of physical models, i.e. the relation of frequencies and intensities of spectral lines to molecular motions of various types. As the precision of measurements made with various experimental techniques increases relentlessly, the interpretation of observed spectra becomes correspondingly challenging. This condition stimulates the search for, and development of, innovative methods of investigating vibrational systems for which a conventional description fails. Intuitively, the most natural model of intramolecular motions involves interacting anharmonic oscillations of atomic canters, but this simple physical model lacks a mathematically exact solution. The use of perturbation theory, however, solves the problem. This classical method is simple and clear, but its application is generally limited to the first few orders of theory that any textbook on quantum mechanics describes. The determination of corrections of higher orders becomes complicated through the sheer bulk of the calculations. The calculation of frequencies and intensities of spectral lines with an accuracy defined by experiment hence becomes difficult. A real spectrum of a sample containing even diatomic molecules of a particular chemical compound can consist of several thousand lines. Despite these difficulties, some success has been achieved in developing an adequate method of calculation, embracing perturbation theory. In the following sections, we consider the development of techniques of perturbation theory as applied to problems of molecular spectroscopy to calculate the frequencies and intensities of vibration–rotational transitions.

Historically, a quantum-mechanical consideration of the anharmonicity of diatomic molecules began with Dunham's work [1]; deriving matrix elements for vibrational transitions up to the third derivative of the dipolar moment in terms of perturbation theory, he determined a numerical value for the second derivative

Quantum Theory of Anharmonic Effects in Molecules. DOI: http://dx.doi.org/10.1016/B978-0-12-397912-4.00002-7

of the dipolar-moment function of a HCl molecule from the experimental distribution of intensities in the infrared spectrum of a gaseous sample. Using various computational methods and varied initial assumptions about functions for potential energy and dipolar moment, other authors have subsequently tried to improve the techniques of calculations [2–6]. In this regard, we mention specifically the hypervirial theorem [6,7], the method of Feynman diagrams [8] and the canonical or contact transformations [9]. The objective of the respective authors was typically the eventual results; the procedure of the calculations was thus afforded little attention.

Although for diatomic molecules an application of the hypervirial theorem was fruitful in calculations of matrix elements of a one-dimensional anharmonic oscillator through recurrence relations [10,11], this method is inefficient for polyatomic molecules.

The method of Feynman graphs enables one to eliminate the recurrence scheme of perturbation theory. Circumventing calculations of preceding orders, one might work directly with expressions for wave functions and energy of arbitrary order [8]; this capability is a great advantage of this method. A characteristic of problems in molecular spectroscopy is, however, that one must initially calculate corrections of low order and only then proceed to approximations of higher order. The stated advantage for calculations of low order is rapidly lost in corrections of higher order. For example, the conversion of a diagram of twentieth order into an algebraic expression becomes a complicated procedure in which one must be concerned about the risk of error.

A systematic investigation of vibration–rotational spectra of polyatomic molecules has been conducted mainly with the method of contact transformations [3,12–14], which allowed the retention of the q-number approach and eliminated a problem of superfluous summation over the matrix elements. Although corrections in canonical transformation theory are considered to be equivalent to approximations of the common perturbation theory, this point of view is inaccurate; rather, this method can be used to choose an effective Hamiltonian. For instance, Watson [15] proposed a hypothesis that there exist many rotational Hamiltonians, which all describe experimental data equally validly. Choosing an initial Hamiltonian, by means of a convenient canonical transformation, we obtain another Hamiltonian that yields the same eigenvalues and has a simple parameterization for the interpretation of experimental data. The principal deficiency of this method is that it lacks a clear form of all expressions; as a result, formulae become much too bulky and impede a clear understanding by experimenters.

Dunham's practice of standard perturbation theory can be extrapolated to polyatomic molecules, but alternative algorithms of perturbation theory for the pure vibrational problem have been developed [16]. A novel method within a formulation of quantum theory is based on differentiation with respect to coupling parameters [17]; it produces simple and clear equations for matrix elements [16,18]. Essentially a recurrence scheme, it represents a form of solution involving polynomials of quantum numbers. This formalism allows one to generate rules to calculate observable matrix elements, which determine the frequencies and intensities of vibrational transitions. This approach is reminiscent of Feynman diagrams: we calculate all desired polynomials, make convenient tables and then express physical quantities in terms of the polynomial quantities [19].

The principal objective of this formalism is to simplify the traditional perturbation theory. According to this polynomial method, we accrue all advantages and avoid all shortcomings of the preceding techniques. Efficient for both diatomic and polyatomic molecules, this method is free from the problem of superfluous summation. A convenient recurrence scheme implemented with contemporary computers allows one to optimize all calculations and to decrease greatly the duration of calculations of vibrational frequencies and intensities. When we allude here to approximations of higher order, we have in mind perturbation theory in the tenth or twentieth orders.

Introducing this formalism certainly does not solve all problems: many specific questions, such as those concerned with the effects of vibration–rotational interaction, remain. For instance, a theorem of extraneous quantum numbers has been formulated [20]; with its help an exact solution for coefficients of the Herman–Wallis factor has been obtained [18] – this method is highly original. As a result, we greatly simplify the calculation of intensities for diatomic molecules. For arbitrary linear polyatomic molecules, a comparable success is foreseen, but the possibility of extending this theorem to describe the vibration–rotational spectra of non-linear molecules has yet to be investigated.

Frequencies and Intensities

The simplest choice to describe oscillations is a model of harmonic oscillators with frequencies ω_k. In this case, the Hamiltonian has a simple form

$$H^0 = \hbar \sum_{k=1}^{r} \frac{\omega_k}{2}(p_k^2 + q_k^2),$$

in which $\hbar$ is the Planck constant, r is the number of normal vibrations, p_k denotes momentum and q_k normal coordinate. The eigenvalues of H^0 are

$$E_n^0 = \hbar \sum_k \omega_k \left(n_k + \frac{1}{2}\right), \quad n_k = 0, 1, 2, \ldots.$$

Eigenvector $|n_1, n_2, \ldots, n_k, \ldots\rangle$ is a product of individual functions $|n_1\rangle$, $|n_2\rangle, \ldots$ of each oscillator. The effects of anharmonicity are taken into account in terms of perturbation theory for the stationary states of the Hamiltonian of general type:

$$H = H^0 + W,$$

in which perturbation function W represents an expansion in normal coordinates q_k and hence in powers of a small parameter, λ:

$$W = \sum_{p>0} \lambda^p \sum_{(j_1 j_2 \cdots j_r)p+2} a_{j_1 j_2 \cdots j_r} \xi_1^{j_1} \xi_2^{j_2} \cdots \xi_r^{j_r} \equiv \sum_{p>0} \lambda^p G_p,$$

$$\xi_k = \sqrt{2}q_k, \quad k = 1, 2, \ldots, r.$$

Here, $a_{j_1 j_2 \cdots j_r}$ are the anharmonic force coefficients. A special summation is performed over the indices in parentheses; symbol $(j_1\, j_2 \cdots j_r)p+2$ signifies a summation over $j_1, j_2, \ldots, j_r$ under the constraint that

$$j_1 + j_2 + \cdots + j_r = p + 2.$$

In what follows, when using such a summation, we denote the set of indices associated with the vibrational variables, e.g. $j_1, j_2, \ldots, j_r$ as j.

We summarize the basic requirements to the computational formalism of the anharmonicity problem. First, the concurrence of separate orders must be correctly taken into account, and the contribution of each perturbation group G_p to the sought result must be considered. The first perturbation order is determined by the quantity G_1, the second by G_1 and G_2 and so on. Second, advantage must be taken of the recurrent character of perturbation theory, and algebraic expressions for corrections of higher order must be derived from the lowest approximations. This approach allows one to avoid repeated calculations, as information on the perturbation is already involved in the preceding approximation to which there is no need to return. Third, difficulties of renormalization of the wave function when proceeding from a current correction to that of the next order must be overcome. This requirement provides more subtle work with experimental data. Finally, to save all expressions in a clear manner, the final formulae must not be bulky.

Molecular rotation is considered in an analogous manner. Initially, satisfying the model of a rigid rotator, in the Hamiltonian appears expression

$$B_a J_a^2 + B_b J_b^2 + B_c J_c^2,$$

in which J_a, J_b and J_c are total angular momentum components and B_a, B_b and B_c are rotational parameters. The axes of the coordinate system $\{abc\}$ are along the inertial axes of the molecule. If necessary, we include in the Hamiltonian other terms of type $J_\alpha J_\beta$ with $\alpha \neq \beta$, and then terms simultaneously containing p_k, q_ℓ and J_α. As a result, a possible form of a rotational Hamiltonian is

$$\frac{1}{2}\sum_{\alpha\beta} \Theta_{\alpha\beta}(J_\alpha - l_\alpha)(J_\beta - l_\beta),$$

in which $\Theta_{\alpha\beta}$ is an inverse inertial tensor and $l_\alpha = \sum_{kj}\zeta_{kj}^{\alpha} q_k p_j$ are by definition components of vibrational momentum [21,22]. Quantities ζ_{kj}^{α} are called Coriolis coefficients. A description of vibration–rotational interaction is concerned with components l_α and also with the dependence of matrix elements $\Theta_{\alpha\beta}$ on normal coordinates q_k.

The eigenvalues of the Hamiltonian determine the frequencies; the eigenfunctions determine the intensities of vibration–rotational transitions. Line strength

S_{mn} is proportional to the transition energy, $\hbar\omega_{mn}$; apart from that common factor, the type of absorption is determined by the Einstein probabilities, i.e. by the squared matrix elements of electric $\mathbf{d}$ and magnetic $\boldsymbol{\mu}$ dipolar moments, electric quadrupolar moment Q and so on. Essentially, the quantity S_{mn} defines the intensity. We have

$$S_{mn} = \hbar\omega_{mn}(c_1|\mathbf{d}_{mn}|^2 + c_2|\boldsymbol{\mu}_{mn}|^2 + c_3|Q_{mn}|^2 + \cdots).$$

Coefficients c_1, c_2, c_3,... are independently determined for each concrete physical problem.

From a practical point of view, the electric-dipolar transitions are of greatest interest. For example, consider this expression for S_{mn} for free molecules:

$$S_{mn} = \hbar\omega_{mn}\cdot \frac{4\pi^2}{3\hbar c}|(m|\mathbf{d}|n)|^2(1 - \mathrm{e}^{-\hbar\omega_{mn}/k_{\mathrm{B}}T})(Ng_n/Q)\mathrm{e}^{-E_n/k_{\mathrm{B}}T}.$$

Here, all quantities are simply interpreted [6,23]. Specifically,

$$\frac{4\pi^2}{3\hbar c}|(m|\mathbf{d}|n)|^2$$

follows from an expression for the transition probability per second obtained in the first order of perturbation theory. Transition energy $\hbar\omega_{mn}$ is equal to $E_m - E_n$, of which E_n and E_m are the energies of a molecule that belong to eigenstates $|n)$ and $|m)$. Quantity

$$N_n = (Ng_n/Q)\mathrm{e}^{-E_n/k_{\mathrm{B}}T}$$

from the Boltzmann law defines a fraction of molecules in the initial state with energy E_n at temperature T. Here, N is the concentration of molecules, k_{B} is the Boltzmann constant, g_n is the degeneracy of level E_n and Q is the partition function, for which

$$Q = \sum_s g_s\, \mathrm{e}^{-E_s/k_{\mathrm{B}}T}.$$

The factor

$$1 - \mathrm{e}^{-\hbar\omega_{mn}/k_{\mathrm{B}}T} = 1 - N_m g_n/N_n g_m,$$

in which N_m is a number of molecules in the final state with degeneracy g_m, takes into account the effects of induced emission. This factor is generally near unity.

The principal problem of a calculation of intensity is reduced to the calculation of matrix elements of the electric-dipolar moment function between exact eigenfunctions of a molecular Hamiltonian. Although this procedure is complicated, in particular cases it is possible to simplify the problem. For instance, for diatomic molecules one introduces the Herman–Wallis factor,

$$|(nK|d|n'K')|^2 = (n|d|n')^2(1 + C_{nn'}\Delta K + D_{nn'}\Delta K^2 + \cdots), \quad \Delta K = K' - K,$$

in which $K = (1/2)J(J+1)$, J and n are rotational and vibrational quantum numbers, respectively, $(n|d|n')$ is the rotationless matrix element and $C_{nn'}$, $D_{nn'}$, ... are coefficients of the Herman–Wallis factor [24]; vector $|\cdots)$, which is terminated with a parenthesis, characterizes the exact state (see later). A similar expansion is applicable for linear polyatomic molecules. In general, it is not possible to factorize exactly a squared matrix element into vibrational and rotational parts. The problem of a calculation of matrix elements is further complicated in that a correct explanation of spectra must take into account the anharmonicity caused by the non-linearity of the dipolar-moment function. So, for an arbitrary polyatomic molecule, we have

$$d = \sum_{\ell}\sum_{(s)\ell} \frac{2^{-\ell/2}}{\ell!} d^{(\ell)}_{s_1 s_2 \cdots s_r} \xi_1^{s_1} \xi_2^{s_2} \cdots \xi_r^{s_r}.$$

Coefficients $d^{(\ell)}_{s_1 s_2 \cdots s_r}$ in this expansion of the dipolar-moment function in normal coordinates q_k characterize the electro-optical anharmonicity of molecular vibrations. The higher is the overtone, the greater is the influence of the non-linear part of function d on overtone intensity.

Perturbation Algebra

For calculations performed with perturbation theory of only first order, the computational technique used is of little importance. For an order beyond the first, the computational procedure must be modified so that it is clear and convenient for solving the specific problem under consideration.

We begin with a general description. Let a system with Hamiltonian H_0 be subject to perturbation H'. The principal objective of stationary perturbation theory is to find eigenvalues ε and eigenfunctions $|u\rangle$ of Hamiltonian $H_0 + H'$ as expansions in powers of a perturbation having eigenvalues ε_0 and eigenfunctions $|u_0\rangle$ of zero-order Hamiltonian H_0. In essence, this formulation yields recurrence relations for the sought corrections of eigenvalues and eigenfunctions. A direct calculation clarifies the procedure.

Assuming equation $(H_0 + H')|u\rangle = \varepsilon|u\rangle$, into which we substitute

$$\varepsilon = \sum_k \varepsilon_k \text{ and } |u\rangle = \sum_k |u_k\rangle,$$

in which ε_k and $|u_k\rangle$ are corresponding corrections of order k in the perturbation, and comparing quantities of the same order, we obtain this relation of Rayleigh–Schrödinger theory,

$$H_0|u_k\rangle + H'|u_{k-1}\rangle = \sum_{m=0}^{k} \varepsilon_m |u_{k-m}\rangle.$$

With the aid of this equation, we find an arbitrary correction,

$$\varepsilon_k = \langle u_0|(H' - \varepsilon_1)|u_{k-1}\rangle - \sum_{m=2}^{k-1} \varepsilon_m \langle u_0|u_{k-m}\rangle.$$

This expression clearly demonstrates the recurrence character of perturbation theory: the next approximation invariably depends on the preceding one.

One might improve the result obtained [2]. For this purpose, having used relation $\langle u_0|(H' - \varepsilon_1) = -\langle u_1|(H_0 - \varepsilon_0)$, and then the relation of Rayleigh–Schrödinger theory, we represent matrix element $\langle u_0|(H' - \varepsilon_1)|u_{k-1}\rangle$ in a form

$$\langle u_1|(H' - \varepsilon_1)|u_{k-2}\rangle - \sum_{m=2}^{k-1} \varepsilon_m \langle u_1|u_{k-m-1}\rangle.$$

An arbitrary approximation is thus

$$\varepsilon_k = \langle u_1|(H' - \varepsilon_1)|u_{k-2}\rangle - \sum_{m=2}^{k-2} \varepsilon_m (\langle u_1|u_{k-m-1}\rangle + \langle u_0|u_{k-m}\rangle),$$

in which we take into account the trivial expression $\langle u_0|u_1\rangle + \langle u_1|u_0\rangle = 0$. Joining to this expression the recurrence relation

$$\begin{aligned}\langle u_{k-m}|(H' - \varepsilon_1)|u_{m-1}\rangle &= \langle u_{k-m-1}|(H' - \varepsilon_1)|u_m\rangle \\ &+ \sum_{\ell=2}^{m} \varepsilon_\ell \langle u_{k-m}|u_{m-\ell}\rangle - \sum_{\ell=2}^{k-m} \varepsilon_\ell \langle u_{k-m-\ell}|u_m\rangle,\end{aligned}$$

which is proved directly through the Rayleigh–Schrödinger equation, we obtain in a convenient form the result sought. This recurrence relation evidently allows one to express ε_{2k} and ε_{2k+1} through corrections of order k. Using such a method to solve a problem correctly is, however, not always simple and clear, as there is no strong basis to account for the calculation of matrix element $\langle u_1|(H' - \varepsilon_1)|u_{k-2}\rangle$, for instance, being preferable to the calculation of $\langle u_0|(H' - \varepsilon_1)|u_{k-1}\rangle$ and so on.

Expansions of Two Types

We consider a procedure to calculate the corrections to eigenvalues and eigenfunctions in a framework of stationary perturbation theory. Let E_n and $|\psi_n\rangle$ be eigenvalues and eigenvectors, respectively, of the exact Schrödinger equation

$$(H_0 + H')|\psi_n\rangle = E_n|\psi_n\rangle,$$

in which E_n^0 and $|n\rangle$ are eigenvalues and eigenvectors of unperturbed Hamiltonian H_0, respectively. The essence of perturbation theory is that E_n is nearer to E_n^0 than E_{n+1}^0. Therefore,

$$E_n^0 - E_m^0 \gg |\langle n|H'|m\rangle|$$

and

$$\langle n|H'|m\rangle(E_n^0 - E_m^0)^{-1}$$

can serve as an expansion parameter. This perturbation theory is of the first type, called the Rayleigh–Schrödinger formalism. The smallness of this expansion parameter provides asymptotic convergence of the method. The second type, Brillouin–Wigner, differs slightly from the other case. Instead of $E_n^0 - E_m^0$ we here have $E_n - E_m^0$, and the corresponding expansion parameter is

$$\langle n|H'|m\rangle(E_n - E_m^0)^{-1}.$$

This quantity is also small for $m \neq n$, but one must know the value of exact energy E_n.

To obtain E_n and $|\psi_n\rangle$ in the general case, we introduce projection operator $P = |n\rangle\langle n|$ and some function F, the appearance of which becomes clear as given below. One generally takes into account the untraditional condition of normalization,

$$\langle n|\psi_n\rangle = 1.$$

In this case, $P|\psi_n\rangle = |n\rangle$. We rewrite the Schrödinger equation in a form

$$(F - H_0)|\psi_n\rangle = (F - E_n + H')|\psi_n\rangle$$

and separate from it function $|\psi_n\rangle$:

$$|\psi_n\rangle = \frac{1}{F - H_0}(F - E_n + H')|\psi_n\rangle.$$

This expression has meaning only if the denominator is not equal to zero. To eliminate this undefined quantity, we multiply the result obtained by $1 - P$ on the left side, then

$$|\psi_n\rangle - |n\rangle = \frac{1 - P}{F - H_0}(F - E_n + H')|\psi_n\rangle,$$

in which we take into account that

$$(1 - P)|\psi_n\rangle = |\psi_n\rangle - |n\rangle,$$

so that

$$|\psi_n\rangle = \left[1 - \frac{1-P}{F-H_0}(F-E_n+H')\right]^{-1}|n\rangle.$$

To proceed, we use the operator expansion

$$\frac{1}{A-X} = \frac{1}{A} + \frac{1}{A}X\frac{1}{A} + \frac{1}{A}X\frac{1}{A}X\frac{1}{A} + \cdots.$$

Having assumed $A = 1$ and $X = ((1-P)/(F-H_0))(F-E_n+H')$, we have as a result

$$|\psi_n\rangle = \sum_{i=0}^{\infty}\left[\frac{1-P}{F-H_0}(F-E_n+H')\right]^i|n\rangle.$$

Further, as $\langle n|\psi_n\rangle = 1$, from the Schrödinger equation it follows that

$$E_n = E_n^0 + \langle n|H'|\psi_n\rangle.$$

Thus,

$$E_n = E_n^0 + \sum_{i=0}^{\infty}\left\langle n\middle|H'\left[\frac{1-P}{F-H_0}(F-E_n+H')\right]^i\middle|n\right\rangle.$$

General formulae for the energy and for the function in Rayleigh–Schrödinger theory are expressible at $F = E_n^0$. As a result,

$$|\psi_n\rangle = \sum_{i=0}^{\infty}\left[\frac{1-P}{E_n^0-H_0}(E_n^0-E_n+H')\right]^i|n\rangle,$$

$$E_n = E_n^0 + \sum_{i=0}^{\infty}\left\langle n\middle|H'\left[\frac{1-P}{E_n^0-H_0}(E_n^0-E_n+H')\right]^i\middle|n\right\rangle.$$

These expressions determine the sought E_n and $|\psi_n\rangle$ with the desired accuracy in powers of perturbation H'. From a practical point of view, quantity E_n is represented in a form of expansion:

$$E_n = E_n^0 + E_n^1 + E_n^2 + E_n^3 + \cdots,$$

in which E_n^k is the correction of order k to the energy in perturbation H'. We note directly that $E_n^1 = \langle n|H'|n\rangle$, as it should be. For the second-order correction, we have

$$E_n^2 = \left\langle n \middle| H' \left[\frac{1-P}{E_n^0 - H_0} (H' - E_n^1) \right] \middle| n \right\rangle,$$

but

$$\frac{1-P}{E_n^0 - H_0} |n\rangle = 0;$$

therefore,

$$E_n^2 = \langle n | H' \eta^{-1} H' | n \rangle, \quad \eta^{-1} = \frac{1-P}{E_n^0 - H_0}.$$

In an analogous manner, one defines the third-order correction,

$$E_n^3 = \langle n | H' \eta^{-1} H' \eta^{-1} H' | n \rangle - E_n^1 \langle n | H' \eta^{-2} H' | n \rangle.$$

In Brillouin–Wigner theory, $F = E_n$. Consequently,

$$|\psi_n\rangle = \sum_{i=0}^{\infty} \left[\frac{1-P}{E_n - H_0} H' \right]^i |n\rangle \text{ and } E_n = E_n^0 + \sum_{i=0}^{\infty} \left\langle n \middle| H' \left[\frac{1-P}{E_n - H_0} H' \right]^i \middle| n \right\rangle.$$

This formulation is rarely used because one must find the exact solution of the eigenvalue equation for E_n and then determine $|\psi_n\rangle$.

Many-Time Formalism

Let us formulate the principles of time-dependent perturbation theory. One might use a formally non-stationary approach even in those cases in which a perturbation is independent of time. Recalling the previous definitions, we write the Schrödinger equation as

$$\mathrm{i}\hbar \frac{\partial}{\partial t} |u\rangle = (H_0 + H')|u\rangle.$$

Assuming

$$|u\rangle = \mathrm{e}^{-\mathrm{i}H_0 t/\hbar} |\Phi\rangle \text{ and } V = \mathrm{e}^{\mathrm{i}H_0 t/\hbar} H' \, \mathrm{e}^{-\mathrm{i}H_0 t/\hbar},$$

we pass to an interaction picture, then

$$\mathrm{i}\hbar \frac{\partial}{\partial t} |\Phi\rangle = V|\Phi\rangle.$$

We take into account additionally the adiabatic hypothesis, i.e. the perturbation evolves slowly in the interval from $t = -\infty$ until $t = 0$. It is expressible as

$$V_\alpha = e^{iH_0 t/\hbar} H' \, e^{-iH_0 t/\hbar} \, e^{\alpha t},$$

in which $\alpha > 0$. At the end of all calculations α is set equal to zero.

We determine quantity $U_\alpha(t,t_0)$, which, acting on vector $|\Phi(t_0)\rangle$ at the initial moment of time, yields its value at moment t:

$$|\Phi(t)\rangle = U_\alpha(t, t_0)|\Phi(t_0)\rangle.$$

Vector $|\Phi(-\infty)\rangle$ is obviously equal to $|n\rangle$ and $|\Phi(0)\rangle$ is equal to $|\psi_n\rangle$ within the accuracy of the normalization factor.

Now we have an equation for matrix $U_\alpha(t, t_0)$, namely,

$$i\hbar \frac{\partial}{\partial t} U_\alpha(t, t_0) = V_\alpha(t) U_\alpha(t, t_0).$$

From this equation it follows that

$$U_\alpha(t, t_0) = 1 - \frac{i}{\hbar} \int_{t_0}^{t} V_\alpha(t_1) U_\alpha(t_1, t_0) dt_1.$$

Using the method of consecutive iterations, we have

$$U_\alpha(t, t_0) = 1 + \left(-\frac{i}{\hbar}\right) \int_{t_0}^{t} V_\alpha(t_1) dt_1 + \left(-\frac{i}{\hbar}\right)^2 \int_{t_0}^{t} dt_1 \int_{t_0}^{t_1} V_\alpha(t_1) V_\alpha(t_2) U_\alpha(t_2, t_0) dt_2$$

after the first iteration, and

$$U_\alpha(t, t_0) = 1 + \sum_{s=1}^{\infty} U_\alpha^{(s)}(t, t_0)$$

after all iterations. Here,

$$U_\alpha^{(s)}(t, t_0) = \left(-\frac{i}{\hbar}\right)^s \int_{t_0}^{t} dt_1 \int_{t_0}^{t_1} dt_2 \cdots \int_{t_0}^{t_{s-1}} V_\alpha(t_1) V_\alpha(t_2) \cdots V_\alpha(t_s) dt_s,$$
$$\text{at } t_1 > t_2 > \cdots > t_s,$$

or in the Dyson form,

$$U_\alpha^{(s)}(t, t_0) = \left(-\frac{i}{\hbar}\right)^s \frac{1}{s!} \int_{t_0}^{t} dt_1 \int_{t_0}^{t} dt_2 \cdots \int_{t_0}^{t} T[V_\alpha(t_1) V_\alpha(t_2) \cdots V_\alpha(t_s)] dt_s,$$

in which operator T organizes all quantities $V_\alpha(t_k)$ to the right of the values in order of increasing time from right to left. To elucidate the meaning of this expression, we consider $U_\alpha^{(2)}$ in detail, namely,

$$\frac{1}{2!}\int_{t_0}^{t}\mathrm{d}t_1\int_{t_0}^{t}T[V_\alpha(t_1)V_\alpha(t_2)]\mathrm{d}t_2=\frac{1}{2}\int_{t_0}^{t}\mathrm{d}t_1\int_{t_0}^{t_1}V_\alpha(t_1)V_\alpha(t_2)\mathrm{d}t_2$$
$$+\frac{1}{2}\int_{t_0}^{t}\mathrm{d}t_1\int_{t_1}^{t}V_\alpha(t_2)V_\alpha(t_1)\mathrm{d}t_2.$$

In the second integral, we sequentially interchange integration variables by the replacement $t_1\leftrightarrow t_2$, and then the order of integration; as a result,

$$\int_{t_0}^{t}\mathrm{d}t_1\int_{t_1}^{t}V_\alpha(t_2)V_\alpha(t_1)\mathrm{d}t_2=\int_{t_0}^{t}\mathrm{d}t_2\int_{t_2}^{t}V_\alpha(t_1)V_\alpha(t_2)\mathrm{d}t_1=\int_{t_0}^{t}\mathrm{d}t_1\int_{t_0}^{t_1}V_\alpha(t_1)V_\alpha(t_2)\mathrm{d}t_2.$$

Thus,

$$\frac{1}{2!}\int_{t_0}^{t}\mathrm{d}t_1\int_{t_0}^{t}T[V_\alpha(t_1)V_\alpha(t_2)]\mathrm{d}t_2=\int_{t_0}^{t}\mathrm{d}t_1\int_{t_0}^{t_1}V_\alpha(t_1)V_\alpha(t_2)\mathrm{d}t_2.$$

The validity of an analogous expression for $U_\alpha^{(s)}$ is readily provable by induction.

We set t_0 equal to $-\infty$ and t equal to 0, as a result,

$$|\Phi(0)\rangle=U_\alpha(0,-\infty)|n\rangle.$$

To normalize the vector obtained, we recall that the mixed product is here equal to unity, $\langle n|\psi_n\rangle=1$. Therefore,

$$|\psi_n\rangle=\frac{U_\alpha(0,-\infty)|n\rangle}{\langle n|U_\alpha(0,-\infty)|n\rangle},\qquad U_\alpha(0,-\infty)=1+\sum_{s=1}^{\infty}U_\alpha^{(s)}(0,-\infty),$$

$$U_\alpha^{(s)}(0,-\infty)=\left(-\frac{\mathrm{i}}{\hbar}\right)^s\int_{-\infty}^{0}\mathrm{d}t_1\int_{-\infty}^{t_1}\mathrm{d}t_2\cdots\int_{-\infty}^{t_{s-1}}V_\alpha(t_1)V_\alpha(t_2)\cdots V_\alpha(t_s)\mathrm{d}t_s.$$

We join to this result the formula for an energy-level shift $\Delta E_n=E_n-E_n^0=\langle n|W|\psi_n\rangle$, set α equal to zero, and finally obtain

$$|\psi_n\rangle=\lim_{\alpha\to 0}\frac{U_\alpha(0,-\infty)|n\rangle}{\langle n|U_\alpha(0,-\infty)|n\rangle},\qquad \Delta E_n=\lim_{\alpha\to 0}\frac{\langle n|H'U_\alpha(0,-\infty)|n\rangle}{\langle n|U_\alpha(0,-\infty)|n\rangle}.$$

These expressions are called the formulae of Gell-Mann and Low [25].

To illustrate the efficiency of this many-time formalism, we consider a formal solution of a stationary problem in which H' has no explicit dependence on time. We have

$$U_\alpha^{(s)}(0,-\infty)|n\rangle = \left(-\frac{i}{\hbar}\right)^s \int_{-\infty}^{0} dt_1 \int_{-\infty}^{t_1} dt_2 \cdots \int_{-\infty}^{t_{s-1}} V_\alpha(t_1)V_\alpha(t_2)\cdots e^{iH_0t_s/\hbar}H'\,e^{-iH_0t_s/\hbar}\,e^{\alpha t_s}|n\rangle dt_s$$

$$= \left(-\frac{i}{\hbar}\right)^s \int_{-\infty}^{0} dt_1 \int_{-\infty}^{t_1} dt_2 \cdots \int_{-\infty}^{t_{s-1}} V_\alpha(t_1)V_\alpha(t_2)\cdots e^{\alpha t_s + i(H_0-E_n^0)t_s/\hbar}H'|n\rangle dt_s.$$

Integrating over t_s, we find

$$U_\alpha^{(s)}|n\rangle = \left(-\frac{i}{\hbar}\right)^s \int_{-\infty}^{0} dt_1 \int_{-\infty}^{t_1} dt_2 \cdots \int_{-\infty}^{t_{s-2}} V_\alpha(t_1)\cdots V_\alpha(t_{s-1})\frac{e^{\alpha t_{s-1}+i(H_0-E_n^0)t_{s-1}/\hbar}}{\alpha+i(H_0-E_n^0)/\hbar}H'|n\rangle dt_{s-1}.$$

Further,

$$V_\alpha(t_{s-1}) = e^{iH_0t_{s-1}/\hbar}H'\,e^{-iH_0t_{s-1}/\hbar}\,e^{\alpha t_{s-1}},$$

vector $U_\alpha^{(s)}|n\rangle$ is therefore equal to

$$\left(-\frac{i}{\hbar}\right)^s \int_{-\infty}^{0} dt_1 \cdots \int_{-\infty}^{t_{s-2}} V_\alpha(t_1)\cdots e^{2\alpha t_{s-1}+i(H_0-E_n^0)t_{s-1}/\hbar}H'\frac{1}{\alpha+i(H_0-E_n^0)/\hbar}H'|n\rangle dt_{s-1}$$

$$= \left(-\frac{i}{\hbar}\right)^s \int_{-\infty}^{0} dt_1 \cdots \int_{-\infty}^{t_{s-3}} V_\alpha(t_1)\cdots \frac{e^{2\alpha t_{s-2}+i(H_0-E_n^0)t_{s-2}/\hbar}}{2\alpha+i(H_0-E_n^0)/\hbar}H'\frac{1}{\alpha+i(H_0-E_n^0)/\hbar} \times H'|n\rangle dt_{s-2}.$$

Continuing this script, we proceed to the next formula

$$U_\alpha^{(s)}|n\rangle = \frac{1}{is\hbar\alpha+E_n^0-H_0}H'\frac{1}{i(s-1)\hbar\alpha+E_n^0-H_0}H'\cdots\frac{1}{i\hbar\alpha+E_n^0-H_0}H'|n\rangle,$$

which contains s times operator H'. This result is amazing: it clearly yields the contribution of order s in a perturbation to matrix U, and has no simultaneous clear physical meaning. Each denominator in the sum of this expression is equal to zero as $\alpha\to 0$ when the intermediate state coincides with state $|n\rangle$. It is intuitively clear that divergent contributions lacking a physical meaning should be omitted. In the following section, we solve this problem completely, so that the correctness of this solution becomes recognized.

Methods of Quantum-Field Theory

In contrast with the triviality of perturbation theory in quantum mechanics of one body, the problem of interaction of many bodies demands the involvement of more powerful computational methods. Success in this respect has been achieved in a description of collective motions in terms of quantum-field theory; notable examples are the Brueckner theory of nuclear matter [26,27] and the Goldstone formalism applied to a system of interacting fermions [28]. To elucidate the meaning of quantum-field methods, we consider a system of many fermions with an instantaneous two-body potential $W(\mathbf{r}-\mathbf{r}')$, in which $\mathbf{r}$ and $\mathbf{r}'$ represent current radius vectors of arbitrary particles [29]. Let the contribution of energy W into system Hamiltonian H be H', then

$$H = H_0 + H'.$$

Quantity H_0 represents the sum of one-particle Hamiltonians $H_0^{(k)}$, each of which includes the kinetic and potential energies of a separate particle. Moreover, one-particle eigenfunctions ψ_k and their eigenvalues $E_0^{(k)}$, which follow from equation

$$H_0^{(k)}\psi_k = E_0^{(k)}\psi_k,$$

are known.

According to the general ideology of the second-quantized formalism, we introduce particle-field operators as

$$\varphi(\mathbf{r}) = \sum_k c_k\psi_k(\mathbf{r}) \text{ and } \varphi^+(\mathbf{r}) = \sum_k \psi_k^*(\mathbf{r})c_k^+,$$

with these conventional anti-commutation relations:

$$[\varphi(\mathbf{r}),\varphi^+(\mathbf{r}')]_+ \equiv \varphi\varphi^+ + \varphi^+\varphi = \delta(\mathbf{r}-\mathbf{r}'),$$

$$[\varphi(\mathbf{r}),\varphi(\mathbf{r}')]_+ = [\varphi^+(\mathbf{r}),\varphi^+(\mathbf{r}')]_+ = 0,$$

$$[c_k,c_j^+]_+ = \delta_{kj}, [c_k,c_j]_+ = [c_k^+,c_j^+]_+ = 0.$$

Quantities c_k and c_k^+ are operators independent on time. The Hamiltonian in zero-order approximation in terms of these variables has a form

$$H_0 = \sum_k E_0^{(k)}\nu_k, \quad \nu_k = c_k^+c_k,$$

in which ν_k is an operator of particle number with eigenvalues 1 and 0. In an analogous manner, we write the interaction Hamiltonian:

$$H' = \frac{1}{2}\int\int \varphi^+(\mathbf{r}')\varphi(\mathbf{r}')W(\mathbf{r}-\mathbf{r}')\varphi^+(\mathbf{r})\varphi(\mathbf{r})d\mathbf{r}\, d\mathbf{r}' = \frac{1}{2}\sum_{kjmn}\langle kj|W|mn\rangle c_k^+c_j^+c_mc_n,$$

$$\langle kj|W|mn\rangle = \iint \psi_k^*(\mathbf{r})\psi_j^*(\mathbf{r}')W(\mathbf{r}-\mathbf{r}')\psi_m(\mathbf{r})\psi_n(\mathbf{r}')\mathrm{d}\mathbf{r}\,\mathrm{d}\mathbf{r}',$$

in which factor half indicates that each interaction is taken into account only once.

We introduce the interaction picture. Understanding that

$$\mathrm{e}^{\mathrm{i}H_0t/\hbar}c_k^+c_j\cdots\mathrm{e}^{-\mathrm{i}H_0t/\hbar} = (\mathrm{e}^{\mathrm{i}H_0t/\hbar}c_k^+\,\mathrm{e}^{-\mathrm{i}H_0t/\hbar})\cdot(\mathrm{e}^{\mathrm{i}H_0t/\hbar}c_j\,\mathrm{e}^{-\mathrm{i}H_0t/\hbar})\cdot\cdots,$$

we have

$$V_\alpha(t) = \mathrm{e}^{\mathrm{i}H_0t/\hbar}H'\,\mathrm{e}^{-\mathrm{i}H_0t/\hbar}\,\mathrm{e}^{\alpha t} = \frac{1}{2}\sum_{kjmn}\langle kj|W|mn\rangle c_k^+(t)c_j^+(t)c_m(t)c_n(t)\mathrm{e}^{\alpha t},$$

in which

$$c_n(t) = \mathrm{e}^{\mathrm{i}H_0t/\hbar}c_n\,\mathrm{e}^{-\mathrm{i}H_0t/\hbar} = \mathrm{e}^{-\mathrm{i}E_0^{(n)}t/\hbar}c_n \quad\text{and}\quad c_n^+(t) = \mathrm{e}^{\mathrm{i}E_0^{(n)}t/\hbar}c_n^+$$

through

$$\begin{aligned}c_n(t) &= \mathrm{e}^{\mathrm{i}H_0t/\hbar}c_n\,\mathrm{e}^{-\mathrm{i}H_0t/\hbar} = \mathrm{e}^{\mathrm{i}H_0t/\hbar}c_n\left(\sum_s\frac{1}{s!}\left(-\frac{\mathrm{i}t}{\hbar}\right)^s(H_0)^s\right)\\ &= \mathrm{e}^{\mathrm{i}H_0t/\hbar}\left(\sum_s\frac{1}{s!}\left(-\frac{\mathrm{i}t}{\hbar}\right)^s\left(H_0+E_0^{(n)}\right)^s\right)c_n = \mathrm{e}^{\mathrm{i}H_0t/\hbar}\,\mathrm{e}^{-\mathrm{i}(H_0+E_0^{(n)})t/\hbar}c_n\\ &= \mathrm{e}^{-\mathrm{i}E_0^{(n)}t/\hbar}c_n.\end{aligned}$$

During this calculation, we use an elementary commutator,

$$[c_n, H_0] = \sum_{n'}E_0^{(n')}[c_n, c_{n'}^+c_{n'}] = \sum_{n'}E_0^{(n')}[c_n, c_{n'}^+]_+c_{n'} = E_0^{(n)}c_n,$$

from which $c_nH_0 = (H_0 + E_0^{(n)})c_n$.

We remind further of the Dyson expression for the U-matrix obtained in the preceding section,

$$U_\alpha(t,-\infty) = 1+\sum_{s=1}^{\infty}\left(-\frac{\mathrm{i}}{\hbar}\right)^s\frac{1}{s!}\int_{-\infty}^{t}\mathrm{d}t_1\int_{-\infty}^{t}\mathrm{d}t_2\cdots\int_{-\infty}^{t}T[V_\alpha(t_1)V_\alpha(t_2)\cdots V_\alpha(t_s)]\mathrm{d}t_s.$$

The calculation of this matrix in an explicit form allows one to determine the wave function and energy of the ground state of the system under consideration. We substitute here quantities $V_\alpha(t)$, which we express in terms of field operators,

$$\varphi(x) = \sum_k c_k(t)\psi_k(\mathbf{r}) \quad\text{and}\quad \varphi^+(x) = \sum_k \psi_k^*(\mathbf{r})c_k^+(t),$$

in which, for the purpose of using a diagram technique, we introduce a four-dimensional definition x for t and $\mathbf{r}$. Moreover, we redefine the interaction energy as

$$W_\alpha(x - x') = W(\mathbf{r} - \mathbf{r}')\delta(t - t')\mathrm{e}^{\alpha t}.$$

Through an artificial introduction of the Dirac delta function, in a general expression for the U-matrix, there appears a trivial integration over t'. The final result is obviously

$$U_\alpha(t, -\infty) = 1 + \sum_{s=1}^{\infty}\left(-\frac{\mathrm{i}}{2\hbar}\right)^s \frac{1}{s!}\int \mathrm{d}^4x_1 \int \mathrm{d}^4x_2 \cdots \int \mathrm{d}^4x_s \int \mathrm{d}^4x_1' \cdots \int \mathrm{d}^4x_s'$$
$$\cdot\, T[\varphi^+(x_1')\varphi(x_1')W_\alpha(x_1 - x_1')\varphi^+(x_1)\varphi(x_1)\cdots\varphi^+(x_s')\varphi(x_s')W_\alpha(x_s - x_s')\varphi^+(x_s)\varphi(x_s)].$$

Each integral here implies an integration over all that particular part of space–time behind the surface t.

We derive a mathematical solution. According to the classic paper of Hubbard [30], we try to restore the physical order. We initially resolve our current variables as

$$\varphi(x) = \sum_k^{\text{unocc.}} c_k(t)\psi_k(\mathbf{r}) + \sum_k^{\text{occ.}} c_k(t)\psi_k(\mathbf{r}) \equiv \varphi_{(-)}(x) + \varphi_{(+)}(x)$$

and

$$\varphi^+(x) = \sum_k^{\text{occ.}} \psi_k^*(\mathbf{r})c_k^+(t) + \sum_k^{\text{unocc.}} \psi_k^*(\mathbf{r})c_k^+(t) \equiv \varphi_{(-)}^+(x) + \varphi_{(+)}^+(x),$$

in which $\varphi_{(-)}(x)$ and $\varphi_{(+)}^+(x)$ are correspondingly destruction and creation operators of particles, and $\varphi_{(-)}^+(x)$ and $\varphi_{(+)}(x)$ are destruction and creation operators of holes. Hence, $\varphi_{(+)}^+(x) = [\varphi_{(-)}(x)]^+$ and $\varphi_{(-)}^+(x) = [\varphi_{(+)}(x)]^+$. Goldstone borrowed this terminology of particles and holes directly from positron theory.

According to the Dirac theory, the ground state is a physical vacuum in which no particles exist. Acting with a creation operator on the ground-state vector, one might obtain an arbitrary vector with particles of varied number. The symmetry between electrons and positrons is important. In the Goldstone theory, the ground state is given by an eigenfunction of H_0, i.e. Ψ_0, which necessarily describes a non-degenerate state. In contrast to the electron–positron vacuum, some one-particle states are occupied here. We operate on particles in states, which were unoccupied in initial state Ψ_0 of the system, and holes that were occupied in Ψ_0. The problem is to find the temporal variation of the system from an initial state, which is the ground state by definition. Thus,

$$\varphi_{(-)}(x)|\Psi_0\rangle = 0 = \varphi_{(-)}^+(x)|\Psi_0\rangle,$$

in which the summation for the left side is over unoccupied states and over occupied states for the right side; both occupied and unoccupied states cannot exist in Ψ_0. In this description, an asymmetry exists between particles and holes, but this approach is physically justified because we treat particles of real number. Simply, an unoccupied state becomes occupied and an occupied state becomes unoccupied.

Let us consider the result of the action of the U-matrix on the ground state Ψ_0. In essence, we have destruction and creation field operators of particles and holes in various combinations before Ψ_0. Using the anti-commutation relations, we transfer our variables in such a way that, in the part of expression, all operators $\varphi_{(+)}(x)$ and $\varphi^+_{(+)}(x)$ are to the left of all operators $\varphi_{(-)}(x)$ and $\varphi^+_{(-)}(x)$. Acting on Ψ_0, the destruction operators give zero. As a result, all terms involving $\varphi_{(-)}(x)$ and $\varphi^+_{(-)}(x)$ are omitted, but results after using the anti-commutation relations remain, namely,

$$\begin{aligned}[\varphi_{(+)}(x), \varphi^+_{(-)}(x')]_+ &= \sum_k^{\text{occ.}}\sum_j^{\text{occ.}} \psi_k(\mathbf{r})\psi_j^*(\mathbf{r}')[c_k, c_j^+]_+ e^{i(E_0^{(j)}t' - E_0^{(k)}t)/\hbar} \\ &= \sum_k^{\text{occ.}} \psi_k^*(\mathbf{r}')\psi_k(\mathbf{r})e^{iE_0^{(k)}(t'-t)/\hbar} = \sum_k \nu_k \psi_k^*(\mathbf{r}')\psi_k(\mathbf{r})e^{iE_0^{(k)}(t'-t)/\hbar},\end{aligned}$$

in which to reduce to the latter equality we take into account that in Fermi–Dirac statistics $\nu_k = 1$ for an occupied state and $\nu_k = 0$ for an unoccupied state. In an analogous manner, anti-commutator $[\varphi_{(-)}(x), \varphi^+_{(+)}(x')]_+$ equals

$$\sum_k^{\text{unocc.}} \psi_k^*(\mathbf{r}')\psi_k(\mathbf{r})e^{iE_0^{(k)}(t'-t)/\hbar} = \sum_k (1-\nu_k)\psi_k^*(\mathbf{r}')\psi_k(\mathbf{r})e^{iE_0^{(k)}(t'-t)/\hbar}.$$

We have, however, not various products of field variables but their chronologically ordered combinations. Let us understand what changes. Since in anti-commutators all operators appear in pairs, it suffices to consider the chronological ordering of two operators $\varphi(x)$ and $\varphi^+(x')$. Let $t > t'$ and then for action on Ψ_0, we have

$$T[\varphi(x)\varphi^+(x')]|\Psi_0\rangle = \varphi(x)\varphi^+(x')|\Psi_0\rangle = \sum_j \sum_k^{\text{unocc.}} \psi_j(\mathbf{r})\psi_k^*(\mathbf{r}')e^{i(E_0^{(k)}t' - E_0^{(j)}t)/\hbar} c_j c_k^+ |\Psi_0\rangle.$$

As

$$c_j c_k^+ = \delta_{jk} - c_k^+ c_j,$$

therefore,

$$\begin{aligned}T[\varphi(x)\varphi^+(x')]|\Psi_0\rangle &= \sum_k^{\text{unocc.}} \psi_k(\mathbf{r})\psi_k^*(\mathbf{r}')e^{iE_0^{(k)}(t'-t)/\hbar}|\Psi_0\rangle \\ &\quad - \textstyle\sum_j \displaystyle\sum_k^{\text{unocc.}} \psi_j(\mathbf{r})\psi_k^*(\mathbf{r}')e^{i(E_0^{(k)}t' - E_0^{(j)}t)/\hbar} c_k^+ c_j |\Psi_0\rangle.\end{aligned}$$

The second term here equals

$$-\varphi^+_{(+)}(x')\varphi_{(+)}(x)|\Psi_0\rangle$$

and corresponds to a normal product involving only creation operators. The first term is obtained after calculation of the anti-commutator; it is convenient to represent it as

$$\langle\Psi_0|T[\varphi(x)\varphi^+(x')]|\Psi_0\rangle = \sum_k^{\text{unocc.}} \psi_k(\mathbf{r})\psi_k^*(\mathbf{r}')e^{iE_0^{(k)}(t'-t)/\hbar}.$$

If $t<t'$, then

$$\langle\Psi_0|T[\varphi(x)\varphi^+(x')]|\Psi_0\rangle = -\langle\Psi_0|\varphi^+(x')\varphi(x)|\Psi_0\rangle = -\sum_k^{\text{occ.}} \psi_k(\mathbf{r})\psi_k^*(\mathbf{r}')e^{iE_0^{(k)}(t'-t)/\hbar},$$

in which a minus sign appears through anti-commutation relations of various Fermi operators when $\varphi(x)$ and $\varphi^+(x')$ are interchanged.

To discuss the physical meaning of the diagonal matrix element of a chronologically ordered field-operator pair, we assume that

$$G(x,x') = \langle\Psi_0|T[\varphi(x)\varphi^+(x')]|\Psi_0\rangle = \begin{cases} \langle\Psi_0|\varphi(x)\varphi^+(x')|\Psi_0\rangle, & t>t', \\ -\langle\Psi_0|\varphi^+(x')\varphi(x)|\Psi_0\rangle, & t<t'. \end{cases}$$

In general, this propagator specifies the probability of the corresponding transition of a system; it is sometimes called a Green's function and it is accurate within a factor of $-i$ [31]. For $t>t'$, operator $\varphi^+(x')$ creates a particle in a point of space-time x', which is a point with radius vector $\mathbf{r}'$ at time t'. Then at point x, operator $\varphi(x)$ destroys this particle and the system returns to the initial state. In the second case, when $t<t'$, at point x a hole is created, which then disappears at point x'. If $x=x'$, i.e. $\mathbf{r}=\mathbf{r}'$ and $t'=t+0$, the propagator coincides with density function $\rho(\mathbf{r})$, apart from a sign:

$$G(\mathbf{r},t;\mathbf{r},t+0) = -\sum_k^{\text{occ.}} \psi_k^*(\mathbf{r})\psi_k(\mathbf{r}) = -\rho(\mathbf{r}).$$

What have we derived after all these innovations? In a varied manner, the creation operators $\varphi_{(+)}(x)$ and $\varphi^+_{(+)}(x)$, s operators of interaction W_α, propagators G and a constant coefficient appear in $U_\alpha^{(s)}|\Psi_0\rangle$. The problem is solved, generally speaking. How does the result look? Certainly, one might diligently follow the script from the beginning, and a result must appear. One condition is essential: all quantities mentioned must be composed in all possible ways; through their uniformity the problem becomes greatly simplified. To illustrate the process, we introduce a graphical representation of our quantities:

1. Points of a diagram correspond to arguments x_k and x_k'; one must integrate with respect to all variables x_k and x_k' under conditions that $t_1, t_2, \ldots, t_s, t_1', t_2', \ldots, t_s' < t$;

2. A pair of points x_k and x'_k is joined with a broken line indicating interaction $W_\alpha(x_k - x'_k)$;
3. Operator $\varphi^{+}_{(+)}(x_k)$ is a solid directed line, which runs from a point x_k to the edge of the diagram; for the same particle line running from the edge of the diagram to a point x_k, we introduce operator $\varphi_{(+)}(x_k)$ into the integrand;
4. A solid directed line running from a point x_j to a point x_k corresponds to propagator $G(x_k, x_j)$;
5. Quantity $G(x_k, x_k) = -\rho(\mathbf{r}_k)$ is marked as a loop, in which a particle line runs from a point x_k to itself;
6. The constant coefficient that equals $(2i\hbar)^{-s}(s!)^{-1}$ has no image in the diagram.

Hubbard borrowed these compact rules, first formulated by Feynman, from quantum electrodynamics to describe collective motions in terms of many-body perturbation theory. As a result, each graph implies an algebraic element of the U-matrix.

Diagrams and Computational Rules

We consider a general example to construct an algebraic expression of a diagram. Let a diagram be as shown in Figure 2.1.

The corresponding contribution of sixth order into the U-matrix equals

$$(2i\hbar)^{-6}(6!)^{-1}\int d^4x_1\int d^4x_2\cdots\int d^4x_6\int d^4x'_1\int d^4x'_2\cdots\int d^4x'_6\, W_\alpha(x_1 - x'_1)W_\alpha(x_2 - x'_2)$$

$$\cdots W_\alpha(x_6 - x'_6)\varphi^{+}_{(+)}(x_5)\varphi_{(+)}(x_1)\varphi^{+}_{(+)}(x'_6)\varphi_{(+)}(x'_1)G(x_5, x'_5)G(x'_5, x_4)G(x_4, x'_2)$$

$$\cdot\, G(x'_2, x_1)G(x'_6, x_6)G(x_6, x'_4)G(x'_4, x_3)G(x_3, x'_1)(-\rho(\mathbf{r}_2))(-\rho(\mathbf{r}'_3)).$$

We see that on making a permutation of the labels $(x_k \leftrightarrow x'_k)$, which belong to one interaction line, the result of the integrand becomes the same. Moreover, the result remains the same if we interchange interaction lines, for instance, $x_1x'_1 \leftrightarrow x_4x'_4$. According to Hubbard, we classify the diagrams according to their structures.

We determine class Γ as involving all diagrams that have identical location of the points, the interaction lines and the particle lines. How many diagrams belong to a given class? Let us have a diagram with s interaction lines. The number of

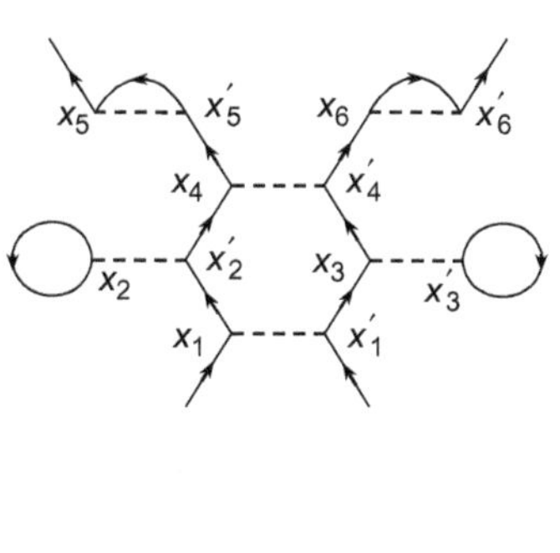

Figure 2.1 Example of diagram structure.

Figure 2.2 Topologically equivalent graphs.

possible interchanges of all interaction lines is obviously equal to $s!$. The number of possible permutations of the points when interaction lines are fixed equals 2^s. The diagrams thus number $2^s s!$ in total. Among these diagrams one might, however, find topologically equivalent structures that have identical integrands, which differ by only a permutation of the variables of integration (Figure 2.2). In each class, one must eliminate equivalent diagrams. As a result, the contribution of diagrams of class Γ becomes

$$U_{\alpha\Gamma} = \frac{2^s s!}{g(\Gamma)} U_{\alpha D},$$

in which $U_{\alpha D}$ is the contribution of only one diagram in class Γ and $g(\Gamma)$ is the number of equivalent diagrams in this class.

Diagrams can be both linked and unlinked. In the first case, one might draw a continuous line through all elements of a diagram. Instances of these structures appear in Figures 2.1 and 2.2. In the second case, we cannot do the same: the unlinked structure clearly consists of separate linked clusters. If a diagram has no external lines running from or to the edge of the diagram, this structure is appropriately called a vacuum diagram.

The U-matrix eventually includes the expansions in various powers of field operators. One might conveniently organize these operators in a form of normally ordered products, each of which can appear in any order of perturbation theory. We rewrite the expression for U_α as an expansion in normal products:

$$U_\alpha = \sum_{N=0}^{\infty} U_{\alpha(N)}.$$

The integrand of $U_{\alpha(N)}$ involves this normal product,

$$[\varphi^+(x_1)\cdots\varphi^+(x_N)\varphi(x_N')\cdots\varphi(x_1')];$$

therefore,

$$\langle\Psi_0|U_\alpha|\Psi_0\rangle = U_{\alpha(0)}.$$

We see that the vacuum expectation value consists of contributions that have no field operator; because external lines correspond to field operators, these diagrams are called vacuum structures. Schematic images of various diagrams are shown in Figure 2.3.

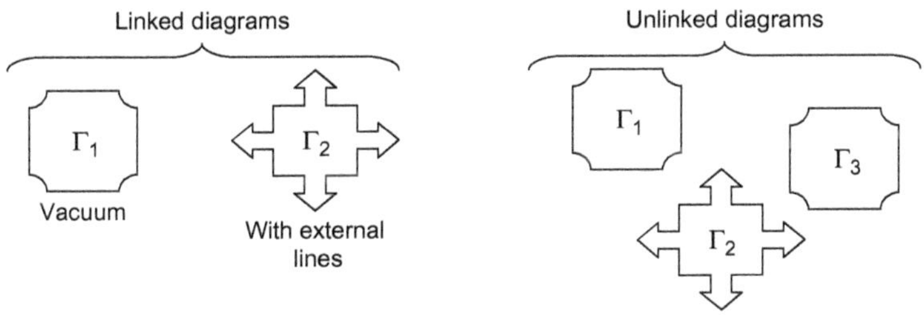

Figure 2.3 Schematic images of linked and unlinked diagrams.

We consider an unlinked structure of class Γ, which comprises p_1 linked structures Γ_1, p_2 linked structures Γ_2, etc. As separate structures $\Gamma_1, \Gamma_2, \ldots$ are not linked to each other with interaction and particle lines, quantity $U_{\alpha\Gamma}$ might be factorized on integration, so that

$$U_{\alpha\Gamma} = \frac{1}{p_1!p_2!\cdots}(U_{\alpha\Gamma_1})^{p_1}(U_{\alpha\Gamma_2})^{p_2}\cdots.$$

On making mutual exchanges of variables that belong to the same linked structures, the integrand of the U-matrix is invariant. We have generally $p_1!$ such exchanges for Γ_1, $p_2!$ for Γ_2, etc. To eliminate extra terms that do not appear in the U-matrix, we introduce an additional factor $(p_1!p_2!\cdots)^{-1}$.

We proceed to calculate the contribution of all possible structures Γ, i.e. quantity U_α. The sums in the expression for $U_{\alpha\Gamma}$ over all quantities $p_1, p_2, \ldots$ run over all integers. We obtain

$$U_\alpha = \sum_{p_1=0}^{\infty}\sum_{p_2=0}^{\infty}\cdots\frac{1}{p_1!p_2!\cdots}(U_{\alpha\Gamma_1})^{p_1}(U_{\alpha\Gamma_2})^{p_2}\cdots = \mathrm{e}^{U_{\alpha\Gamma_1}}\,\mathrm{e}^{U_{\alpha\Gamma_2}}\cdots = \mathrm{e}^{\sum_i U_{\alpha\Gamma_i}}.$$

This result is notable in that the U-matrix is defined as a sum of only linked diagrams [29,30].

It is convenient to resolve the total diagram contribution $U_{\alpha[L]} = \sum_i U_{\alpha\Gamma_i}$ into two parts,

$$U_{\alpha[L]} = U_{\alpha[L_0]} + U_{\alpha[L']},$$

in which $U_{\alpha[L_0]}$ contains all contributions from all vacuum-linked clusters and $U_{\alpha[L']}$ from all linked structures with external lines. Then

$$U_\alpha = \exp(U_{\alpha[L_0]})\exp(U_{\alpha[L']});$$

according to the formulae of Gell-Mann and Low, the ground-state vector is

$$|\Psi\rangle = \lim_{\alpha\to 0}\frac{\exp(U_{\alpha[L_0]})\exp(U_{\alpha[L']})|\Psi_0\rangle}{\langle\Psi_0|U_\alpha|\Psi_0\rangle}.$$

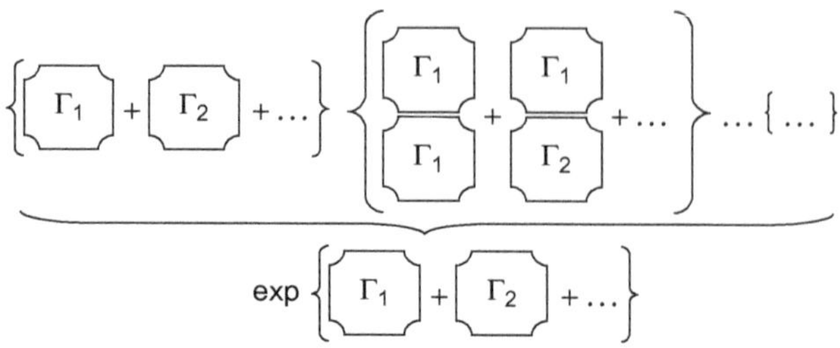

Figure 2.4 Exponent with contributions from only linked structures.

Considering all possible contributions in powers of a perturbation, we understand that each linked cluster that is an unlinked part is also a term of the U-matrix in an expansion with perturbation theory. This term must therefore appear in all orders as the analytical element component. On integration of $U_\alpha|\Psi_0\rangle$ in all variables of time, the contribution from the vacuum part becomes factorized. This factor, determined by all linked and unlinked vacuum diagrams, is expressible in a form of exponent with contributions from only linked diagrams (Figure 2.4) [31,32].

The diagrams with external lines make no contribution to expectation value $\langle\Psi_0|U_\alpha|\Psi_0\rangle$; thus,

$$\langle\Psi_0|U_\alpha|\Psi_0\rangle = \exp(\text{sum of linked vacuum diagrams}).$$

Hence,

$$|\Psi\rangle = \lim_{\alpha\to 0}\frac{\exp(U_{\alpha[L_0]})\exp(U_{\alpha[L']})|\Psi_0\rangle}{\langle\Psi_0|U_\alpha|\Psi_0\rangle} \to \exp(U_{\alpha\to 0[L']})|\Psi_0\rangle.$$

Through the condition that, being a c-number, $U_{\alpha[L_0]}$ includes no field operator, one might factorize and cancel the contribution from the vacuum diagrams in the ground-state wave function.

We might directly take into account the condition to sum over only linked diagrams. According to Goldstone [28], vector $|\Psi\rangle$ simply becomes equal to $U_\alpha|\Psi_0\rangle$ (see also Ref. [29]). Having recalled the formal solution for the wave function obtained above (see section 'Perturbation Algebra'), we have

$$|n\rangle \to |\Psi_0\rangle, \quad |\psi_n\rangle \to |\Psi\rangle, \quad E_n^0 \to E_0,$$

$$U_\alpha(0, -\infty) = 1 + \sum_{s=1}^{\infty} U_\alpha^{(s)}(0, -\infty),$$

$$U_\alpha^{(s)}|\Psi_0\rangle = \frac{1}{\mathrm{i}s\hbar\alpha + E_0 - H_0}H'\frac{1}{\mathrm{i}(s-1)\hbar\alpha + E_0 - H_0}H'\cdots\frac{1}{\mathrm{i}\hbar\alpha + E_0 - H_0}H'|\Psi_0\rangle.$$

This expression acquires a physical meaning as $\alpha\to 0$ if we restrict the summation over all linked diagrams, i.e.

$$|\Psi\rangle = \sum_{s[L']} \left(\frac{1}{E_0 - H_0} H' \right)^s |\Psi_0\rangle.$$

In this sum, function $|\Psi_0\rangle$ cannot appear as an intermediate state, as otherwise the diagram that is excluded must have been unlinked. No denominator thus equals zero as $\alpha \to 0$ and quantity $U_{\alpha[L']}$ is continuous.

The exact energy shift of the ground state is given by

$$\Delta E = E - E_0 = \langle \Psi_0 | H' | \Psi \rangle.$$

Function $|\Psi\rangle$ is hence defined by linked diagrams with external lines. On acting with perturbation operator H', the system returns to the ground state; the diagrams with external lines are automatically omitted. As a result, only vacuum diagrams define a contribution into the energy shift sought:

$$\Delta E = \left\langle \Psi_0 \middle| \sum_{s[L_0]} H' \left(\frac{1}{E_0 - H_0} H' \right)^s \middle| \Psi_0 \right\rangle.$$

This notable formula of Goldstone has played an important role in forming the diagram technique.

In conclusion, one might choose principal diagrams and make a partial summation of the perturbation series in many cases. The point is that, as some contributions of perturbations in many-body problems diverge, it is necessary to take a partial sum to compensate for those divergent terms. One might state that the objective is to obtain an approximate solution, taking into account the diagrams of only a given type. To illustrate the method of a partial summation, we consider the motion of a particle in a system of interacting fermions. The Goldstone formula was obtained for just this case; with its aid Gell-Mann and Brueckner subsequently evaluated the correlation energy of an electron gas in a metal.

Let a particle in motion interact with an electron gas so that only loop diagrams appear. We recall that, for each particle line, we introduce propagator G, and for an interaction line, perturbation operator W_α. We denote the unperturbed function G through G_0, and the contribution from one loop with an interaction line through Σ_0. Perturbed propagator G is obtained on summation of the loop diagrams of all possible orders (Figure 2.5) [33].

We represent this series in a form

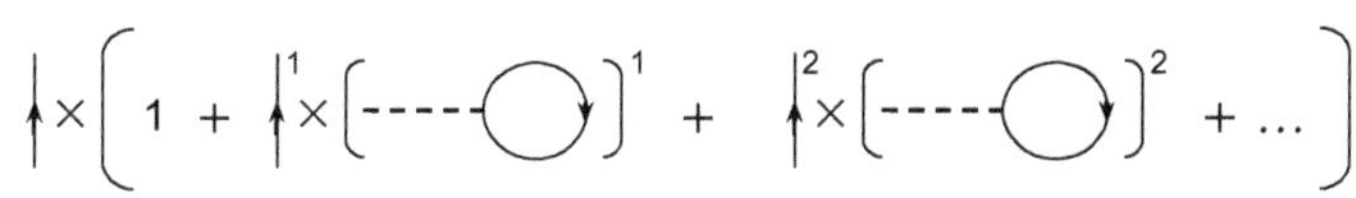

and find its sum. The result sought, which gives quantity G, is

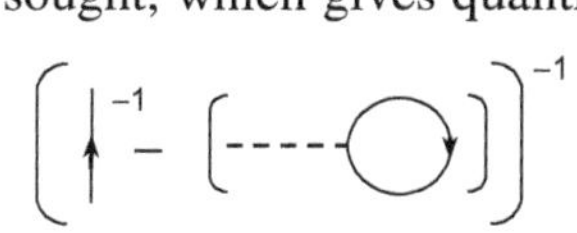

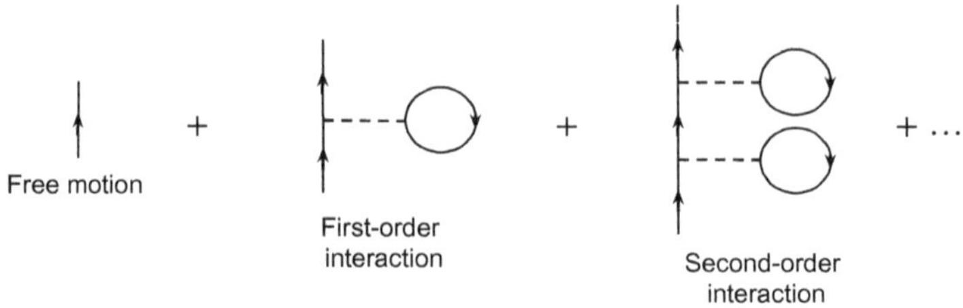

Figure 2.5 Perturbed propagator for loop diagrams.

or

$$G \approx (G_0^{-1} - \Sigma_0)^{-1},$$

in which an approximate equality emphasizes that this propagator is not yet exact because Σ_0 has contributions from only loop diagrams. The simplest loop diagram corresponds to diagonal matrix elements,

$$\begin{aligned}\sum_n^{\text{occ.}} \langle mn|W|mn\rangle &= \sum_n^{\text{occ.}} \int\int \psi_m^*(\mathbf{r})\psi_n^*(\mathbf{r}')W(\mathbf{r}-\mathbf{r}')\psi_m(\mathbf{r})\psi_n(\mathbf{r}')\mathrm{d}\mathbf{r}\,\mathrm{d}\mathbf{r}' \\ &= \int \psi_m^*(\mathbf{r})\left(\int \rho(\mathbf{r}')W(\mathbf{r}-\mathbf{r}')\mathrm{d}\mathbf{r}'\right)\psi_m(\mathbf{r})\mathrm{d}\mathbf{r} = \int \psi_m^*(\mathbf{r})W_{\text{eff}}(\mathbf{r})\psi_m(\mathbf{r})\mathrm{d}\mathbf{r},\end{aligned}$$

in which, according to the rules of 'translation' in the diagrammatic language, density function $\rho(\mathbf{r}')$ here appears. New quantity $W_{\text{eff}}(\mathbf{r})$ represents an effective external potential, in which the particle is multiply scattered. The physical interpretation of this process involves a so-called forward scattering. For instance, we consider the second diagram shown in Figure 2.5, which describes the first-order interaction. The particle initially moves freely, then interacting at a given time; it creates another particle, which immediately disappears at the point of its creation. After that, the particle is in continuous free motion. In an analogous manner, one might interpret the succeeding diagrams, which describe a higher-order interaction.

One can readily guess that an account of loop diagrams corresponds to the well-known Hartree approximation, from which any exchange interaction is excluded. The simplest exchange processes, for instance, are shown in Figure 2.1: they rely on interaction lines x_5x_5' and x_6x_6'. In these cases, creating a new particle, particles have exchanged their places. As a result, the initial particle becomes destroyed and a new particle continues in free motion. The diagram of this type corresponds to the exchange matrix elements:

$$\sum_n^{\text{occ.}} \langle nm|W|mn\rangle = \sum_n^{\text{occ.}} \int\int \psi_n^*(\mathbf{r})\psi_m^*(\mathbf{r}')W(\mathbf{r}-\mathbf{r}')\psi_m(\mathbf{r})\psi_n(\mathbf{r}')\mathrm{d}\mathbf{r}\,\mathrm{d}\mathbf{r}'.$$

Completing the series of loop diagrams with exchange diagrams, one makes a partial summation [33]. In this case, series

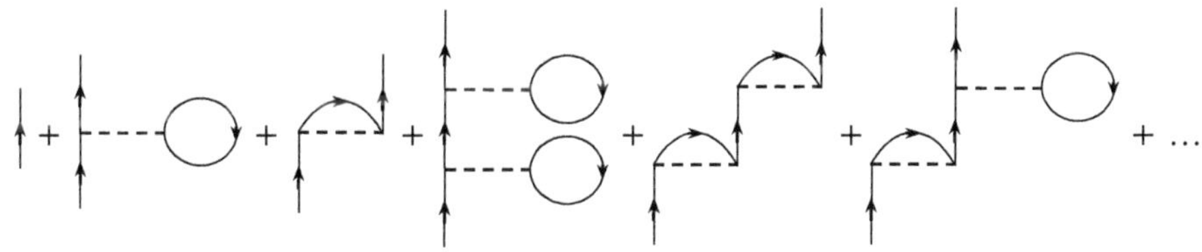

transforms into function

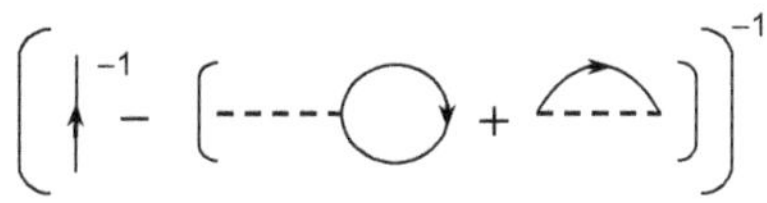

and becomes a description of the Hartree–Fock approximation.

Extending our series with higher-order diagrams, the developed scenario is continued [33,34]. As a result, we obtain the exact equation for a propagator, i.e. the Dyson equation,

$$G = (G_0^{-1} - \Sigma)^{-1} \text{ or } G = G_0 + G_0 \Sigma G,$$

in which Σ is given by a total sum

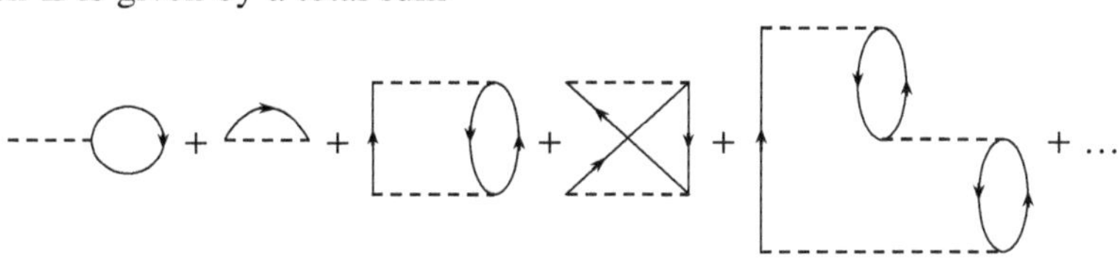

and corresponds to the contribution of irreducible diagrams of the so-called self-energy part. Any diagram with no external lines that can be inserted into the particle line is included in the self-energy part by definition. The simplest examples of these diagrams are loop and exchange structures of various orders considered above. Quantity Σ is sometimes called a mass operator. The meaning of diagrams involved in Σ is simple: these structures are not resolved into parts through the removal of any particle line from the diagram.

It is important to understand that successful work with the diagrams is not the objective of theoretical research. After obtaining physical quantities as functions of the sum from irreducible diagrams, furthermore, one must calculate the contribution from each of those, and only then calculate their total sum. For instance, quantity Σ cannot be practically calculated. The question of choosing principal diagrams, which give the main contribution to matrix elements, remains open. For each concrete problem, for each potential, which plays the role of a perturbation, one must choose, calculate and make a summation with the diagrams determining the principal contribution [31–34]. The diagram ideology, however, implements modern perturbation theory when all the various terms of an arbitrary perturbation can be reduced in accordance with a special 'table' from which it is possible to obtain all observable quantities.

Other Trends and Methods

A calculational technique other than perturbation theory that is widely used is based on a variational principle. With regard to the problems of quantum mechanics for which this principle can be formulated, one must nevertheless numerically diagonalize the Hamiltonian matrix that is preliminarily calculated with the aid of convenient basis functions. These functions might be state vectors for a harmonic or Morse oscillator, or perhaps for a symmetric rotor; in a general case, these functions are products of known vibrational and rotational wave functions. Much here depends on the computing resources and the power of contemporary computers. The variational methods represent a radical direction of investigation and rightly deserve separate consideration. According to our main purpose, we focus here on analytical methods.

Alternative Perturbation Theory

Consider a description of an arbitrary quantum system by means of a dynamical method based on a law of evolution of coupling parameter λ; this parameter characterizes the extent of interaction between particles and fields. By analogy with a Hamiltonian as a shift operator of time, shift operator D acting on the coupling parameter is defined as

$$\mathrm{i}\frac{\mathrm{d}\psi_n}{\mathrm{d}\lambda} = D\psi_n,$$

in which ψ_n is a particular state vector. As

$$-\mathrm{i}\frac{\mathrm{d}\psi_n^*}{\mathrm{d}\lambda} = D\psi_n^*,$$

for an arbitrary physical operator O, we have

$$\frac{\mathrm{d}}{\mathrm{d}\lambda}\langle\psi_m|O|\psi_n\rangle = \langle\psi_m|(\partial O/\partial\lambda + \mathrm{i}[D, O])|\psi_n\rangle.$$

These equations, together with boundary conditions

$$\psi_n = \psi_n^0 \text{ and } O = O^0 \quad \text{as } \lambda = 0,$$

form the basis of a new theory that has as its objective to determine physical quantities taking into account that interaction. The values of these quantities are known at $\lambda = 0$. This formulation of quantum theory is based essentially on differentiation with respect to coupling parameters [17].

This method is applied in quantum-field theory for the determination of scattering matrix S through a differential equation. The famous Dyson formula is readily usable in this description as a formal solution for S. The chronological ordering becomes replaced by ordering in the coupling parameter: according to this theory, one applies differentiation with respect to λ instead of with respect to time. As a result, for stationary problems in quantum mechanics, one derives a significant benefit. For example, if O is the Hamiltonian of system H with eigenvalues E_n and eigenfunctions $|\psi_n\rangle$, we obtain the formula

$$\frac{\mathrm{d}E_n}{\mathrm{d}\lambda} = \left\langle \psi_n \left| \frac{\partial H}{\partial \lambda} \right| \psi_n \right\rangle,$$

which represents an expression of the Hellmann–Feynman theorem [35,36].

These principles show the general trend of the theory. We consider its application to a non-relativistic problem of stationary states. In this case, the Hamiltonian has a simple form

$$H = H_0 + H'.$$

The eigenvalues E_n^0 and eigenfunctions $|n\rangle$ (or ψ_n^0) of H_0 are known as before. Perturbation operator H' includes the coupling parameter λ as a factor,

$$H' = \lambda W.$$

In essence, λ represents a small parameter that characterizes the order of perturbation function W. Using the evolution law on the coupling parameter, one obtains the equations for eigenvalues $E_n(\lambda)$ and eigenfunctions $|\psi_n(\lambda)\rangle$ of perturbed Hamiltonian H.

The approach is formally simple; differentiating a wave equation

$$(H_0 + \lambda W)|\psi_n(\lambda)\rangle = E_n(\lambda)|\psi_n(\lambda)\rangle$$

with respect to λ, we obtain

$$(H_0 + \lambda W)\frac{\mathrm{d}}{\mathrm{d}\lambda}|\psi_n(\lambda)\rangle + W|\psi_n(\lambda)\rangle = \frac{\mathrm{d}E_n(\lambda)}{\mathrm{d}\lambda}|\psi_n(\lambda)\rangle + E_n(\lambda)\frac{\mathrm{d}}{\mathrm{d}\lambda}|\psi_n(\lambda)\rangle.$$

Applying the completeness and orthogonality of eigenfunctions, and taking into account equality $\langle\psi_n(\lambda)|(\mathrm{d}/\mathrm{d}\lambda)|\psi_n(\lambda)\rangle = 0$, which follows from identity

$$\frac{\mathrm{d}}{\mathrm{d}\lambda}\langle\psi_n(\lambda)|\psi_n(\lambda)\rangle = \frac{\mathrm{d}}{\mathrm{d}\lambda}(1) = 0,$$

we represent $\mathrm{d}|\psi_n(\lambda)\rangle/\mathrm{d}\lambda$ as $\sum_{m\neq n} C_{mn}|\psi_m(\lambda)\rangle$, then

$$\sum_{m\neq n} C_{mn}(E_n(\lambda)-E_m(\lambda))|\psi_m(\lambda)\rangle = W|\psi_n(\lambda)\rangle - \frac{\mathrm{d}E_n(\lambda)}{\mathrm{d}\lambda}|\psi_n(\lambda)\rangle.$$

We obtain therefrom this system of exact equations,

$$\frac{\mathrm{d}}{\mathrm{d}\lambda}E_n(\lambda) = \langle\psi_n(\lambda)|W|\psi_n(\lambda)\rangle,$$

$$\frac{\mathrm{d}}{\mathrm{d}\lambda}|\psi_n(\lambda)\rangle = \sum_{m\neq n}\frac{\langle\psi_m(\lambda)|W|\psi_n(\lambda)\rangle}{E_n(\lambda)-E_m(\lambda)}|\psi_m(\lambda)\rangle.$$

The summation is here taken over all states of the perturbed Hamiltonian.

This system of equations is equivalent to the wave equation and demonstrates that the calculations performed in terms of this perturbation theory have a recurrent character. After substituting the series expansions

$$E_n(\lambda) = E_n^0 + \lambda E_n^1 + \lambda^2 E_n^2 + \cdots$$

and

$$\psi_n(\lambda) = \psi_n^0 + \lambda\psi_n^1 + \lambda^2\psi_n^2 + \cdots$$

into the system of equations and comparing quantities of the same order in parameter λ, one finds that, through the first derivative with respect to λ, the corrections on the left side of the equations are one order of magnitude greater than that of the corrections on the right side. For instance,

$$E_n^1 = \langle n|W|n\rangle, \quad |\psi_n^1\rangle = \sum_{m\neq n}\frac{\langle m|W|n\rangle}{E_n^0 - E_m^0}|m\rangle, \ldots.$$

As with solution of any problem in a framework of perturbation theory, one must initially calculate the first-order corrections and only afterwards those of higher order, if required; this method becomes the most appropriate in many cases.

Canonical Transformation

A Hamiltonian of a real system is generally complicated and demands a preliminary simplification of its form. One possibility is a substitution of variables through a convenient canonical transformation. In quantum mechanics, a so-called unitary transformation plays an important role. In this case, arbitrary vector ψ and operator O transform into $\psi_\mathrm{I} = U\psi$ and $O_\mathrm{I} = UOU^{-1}$ correspondingly. Operator U

determining this transformation is subject to an additional condition $U^+ = U^{-1}$. As a result, we obtain new real dynamical variables with the same algebraic relations between them. A well-known historical example of this procedure was concerned with a physically correct interpretation of Dirac matrices $\boldsymbol{\alpha} = (\alpha_1, \alpha_2, \alpha_3)$. We recall that $c\boldsymbol{\alpha}$ represents the velocity operator of a relativistic electron; c is the speed of light in vacuum. Through the non-commutativity of various projections of vector $\boldsymbol{\alpha}$ between themselves, two projections are undefined. One must hence choose another set of coordinates to obtain another representation for Dirac matrices. This problem is eventually reduced to a unitary transformation that resolves the Dirac equation into the Pauli equation, which describes states with positive energy and another equation involving states with negative energy. The resulting transformation is known as the Foldy–Wouthuysen transformation [37].

Despite problems of the physical meaning, the method of unitary transformations has been routinely applied for calculations in terms of perturbation theory. We consider this approach in detail. Hamiltonian H of a system is generally expressible in a form

$$H = H_0 + H' = H_0 + H_1 + H_2 + \cdots,$$

in which H_0 is an unperturbed Hamiltonian with eigenstates $|n\rangle$. Perturbation H' here consists of H_1, H_2, ..., which correspond to perturbations of first order, second order, etc. We transform H with the aid of unitary matrix $U = \exp(\mathrm{i}S)$:

$$H_\mathrm{I} = \mathrm{e}^{\mathrm{i}S} H\, \mathrm{e}^{-\mathrm{i}S}.$$

Operator S conventionally incorporates information about the perturbation. Having expanded exponents in powers of S, we obtain series expansions in perturbation theory. For this purpose, we apply a device. We introduce a function

$$f(x) = \mathrm{e}^{\mathrm{i}xS} H \mathrm{e}^{-\mathrm{i}xS} = \sum_{k=0}^{\infty} \frac{x^k}{k!} f^{(k)}(x)_{x=0},$$

and calculate its derivatives,

$$f'(x)_{x=0} = \mathrm{i}[S, H], \quad f''(x)_{x=0} = \mathrm{i}^2[S, [S, H]],$$

in a general case,

$$f^{(k)}(x)_{x=0} = \mathrm{i}^k \overbrace{[S, [S, \ldots, [S,}^{k} H] \ldots]].$$

We then put $x = 1$ and substitute Hamiltonian H; as a result,

$$H_\mathrm{I} = H + \mathrm{i}[S, H] + \frac{\mathrm{i}^2}{2}[S, [S, H]] + \cdots = H_0^\mathrm{I} + H_1^\mathrm{I} + H_2^\mathrm{I} + \cdots.$$

Here, $H_0^{\mathrm{I}} = H_0$; $H_1^{\mathrm{I}} = H_1 + \mathrm{i}[S, H_0]$ represents the quantity of first order in the perturbation, $H_2^{\mathrm{I}} = H_2 + \mathrm{i}[S, H_1] - (1/2)[S, [S, H_0]]$ the second order, etc. An expression for S is chosen so that all non-diagonal matrix elements of quantity H_1^{I} between unperturbed functions $|n\rangle$ become equal to zero. This theory corresponds to the first order. To obtain the next order, one makes a second unitary transformation $\exp(\mathrm{i}S')$:

$$H_{\mathrm{II}} = H_{\mathrm{I}} + \mathrm{i}[S', H_{\mathrm{I}}] + \frac{\mathrm{i}^2}{2}[S', [S', H_{\mathrm{I}}]] + \cdots = H_0^{\mathrm{II}} + H_1^{\mathrm{II}} + H_2^{\mathrm{II}} + \cdots,$$

and sets equal to zero non-diagonal matrix elements $\langle m|H_2^{\mathrm{II}}|n\rangle$. This general scenario is how one might treat this method of unitary transformation.

This method has had extensive use in the theory of molecular vibrations. Let H_0 be a harmonic Hamiltonian; quantities H_1, H_2,... then represent the terms of an intramolecular potential expansion in normal coordinates q_i; i.e.

$$H = H_0 + H_1 + H_2 + \cdots = H_0 + \sum_{ijk} a_{ijk} q_i q_j q_k + \sum_{ijk\ell} A_{ijk\ell} q_i q_j q_k q_\ell + \cdots,$$

in which a_{ijk}, $A_{ijk\ell}$, ... are anharmonic force parameters. On making a transformation

$$U = \exp(\mathrm{i}S^{(k-1)}) \cdots \exp(\mathrm{i}S')\exp(\mathrm{i}S),$$

Hamiltonian H becomes reduced to diagonal form with an accuracy of order k. Operators S, S', ..., $S^{(k-1)}$, which are called generators, here define the current order of perturbation theory. Quantity S is characterized with coefficients a_{ijk} and determines the first order of theory; S' is characterized by second-order quantities a_{ijk}^2 and $A_{ijk\ell}$ and so on.

Thus reducing the Hamiltonian to a diagonal form, we solve simultaneously the problem of electro-optical anharmonicity – the problem of calculating matrix elements of electric dipolar moment d. Taking into account the first- and higher-order derivatives $d'_i, d''_{ij}, d'''_{ijk}, \ldots$ one represents the dipolar-moment function d of a molecule in a form

$$d = d_0 + d_1 + d_2 + \cdots = d_0 + (1/2)\sum_{ij} d''_{ij} q_i q_j + (1/6)\sum_{ijk} d'''_{ijk} q_i q_j q_k + \cdots,$$

in which $d_0 = d^{(0)} + \sum_i d'_i q_i$ is the dipolar moment in a harmonic approximation. In an analogous manner with a transformation of the Hamiltonian, we calculate quantity UdU^{-1} with the required accuracy. For example, in second order, we have

$$\exp(\mathrm{i}S')\exp(\mathrm{i}S) \approx 1 + \mathrm{i}S + \mathrm{i}S' - (1/2)S^2;$$

after elementary substitutions, we find that $UdU^{-1} \approx d_0^{\mathrm{II}} + d_1^{\mathrm{II}} + d_2^{\mathrm{II}}$, in which

$$d_0^{\mathrm{II}} = d_0,$$

$$d_1^{\mathrm{II}} = d_1 + \mathrm{i}[S, d_0],$$

$$d_2^{\mathrm{II}} = d_2 + \mathrm{i}[S', d_0] + \mathrm{i}[S, d_1] - (1/2)[S, [S, d_0]].$$

Choosing convenient expressions for S and S' from conditions that non-diagonal matrix elements $\langle m|H_1^{\mathrm{I}}|n\rangle$ and $\langle m|H_2^{\mathrm{II}}|n\rangle$ are equal to zero, we readily calculate matrix elements of dipolar moment as $\langle m|UdU^{-1}|n\rangle$ [12].

Hypervirial Result

Apart from general methods characterizing the scheme of real calculations, there have been developed particular methods for specific physical problems. One such development – the method of Feynman graphs – is considered above. With regard to our interest in the problem of anharmonic vibrations, we consider briefly, with an example, a technique known as the hypervirial theorem [7,10,11].

Let a Hamiltonian of a system has a form

$$H = -\partial^2/\partial q^2 + V(q).$$

We calculate a double commutator $[H, [H,\rho]]$ for some function $\rho(q)$:

$$\begin{aligned}
[H,[H,\rho]] &= -[H, \rho'' + 2\rho'\partial/\partial q] \\
&= [H, \rho''] - 2\rho'[H, \partial/\partial q] - 2[H, \rho']\partial/\partial q - 2[H, \rho''] \\
&= -\rho^{\mathrm{IV}} - 2\rho'''\partial/\partial q + 2\rho' V' + 2\rho'''\partial/\partial q + 4\rho''\partial^2/\partial q^2 - 2[H, \rho''] \\
&= -\rho^{\mathrm{IV}} + 2\rho' V' + 4\rho''(V - H) - 2(H\rho'' - \rho'' H) \\
&= -\rho^{\mathrm{IV}} + 2\rho' V' + 4\rho'' V - 2(H\rho'' + \rho'' H).
\end{aligned}$$

For the left side,

$$\langle\psi_n|[H,[H,\rho]]|\psi_{n'}\rangle = (E_{n'} - E_n)^2 \langle\psi_n|\rho|\psi_{n'}\rangle,$$

in which E_n and $|\psi_n\rangle$ are eigenvalues and eigenfunctions of Hamiltonian H.

Putting now $\rho(q) = q^\ell$, one readily obtains an algebraic expression of the hypervirial theorem [11]:

$$\begin{aligned}
(E_{n'} - E_n)^2 \langle\psi_n|q^\ell|\psi_{n'}\rangle = &-\ell(\ell-1)(\ell-2)(\ell-3)\langle\psi_n|q^{\ell-4}|\psi_{n'}\rangle \\
&+ 2\ell\langle\psi_n|q^{\ell-1}V'|\psi_{n'}\rangle + 4\ell(\ell-1)\langle\psi_n|q^{\ell-2}V|\psi_{n'}\rangle \\
&- 2\ell(\ell-1)(E_{n'} + E_n)\langle\psi_n|q^{\ell-2}|\psi_{n'}\rangle.
\end{aligned}$$

One might thereby calculate in a recurrent manner the matrix elements for an anharmonic oscillator and obtain a scheme to determine the exact vibration–rotational energy for a diatomic molecule with Dunham's potential [6].

In particular, the hypervirial theorem includes a diagonal hypervirial result $(n = n')$ [10],

$$\begin{aligned}4\ell(\ell-1)E_n\langle\psi_n|q^{\ell-2}|\psi_n\rangle = &-\ell(\ell-1)(\ell-2)(\ell-3)\langle\psi_n|q^{\ell-4}|\psi_n\rangle\\ &+2\ell\langle\psi_n|q^{\ell-1}V'|\psi_n\rangle+4\ell(\ell-1)\langle\psi_n|q^{\ell-2}V|\psi_n\rangle,\end{aligned}$$

the virial theorem $(n = n'$ and $\ell = 2)$ [38],

$$E_n = \frac{1}{2}\langle\psi_n|qV'|\psi_n\rangle + \langle\psi_n|V|\psi_n\rangle,$$

and Ehrenfest's theorem $(n = n'$ and $\ell = 1)$,

$$\langle\psi_n|V'|\psi_n\rangle = 0,$$

which emphasize its distinctive generality.

This approach has, however, a general deficiency. Even though a problem of an N-dimensional oscillator has high symmetry, the hypervirial theorem is applicable for only a one-dimensional case. Despite this obstacle, there exists a formal extrapolation to a general case in which an effective variable q implies a sum of squared coordinates of an oscillator:

$$q^2 = q_1^2 + q_2^2 + \cdots + q_N^2.$$

To avoid fractional degrees, the potential is defined as

$$V_{\text{eff}}(q) = \sum_{k\geq 0} \upsilon_k q^{2k+2},$$

in which υ_k are coefficients. A calculation of the energy of this isotropic oscillator is simplified through a diagonal hypervirial result [7,39],

$$\langle\psi_n|[q^\ell\partial/\partial q, H_{\text{eff}}]|\psi_n\rangle = \langle\psi_n|q^\ell\partial/\partial q|\psi_n\rangle E_n - E_n\langle\psi_n|q^\ell\partial/\partial q|\psi_n\rangle = 0,$$

in which E_n and $|\psi_n\rangle$ are eigenvalues and eigenfunctions of a new effective Hamiltonian H_{eff}. This Hamiltonian H_{eff} includes effective potential V_{eff} and is generally expressed in terms of hyperspherical coordinates [39]. What do we see? According to this model, in which the influence of anharmonicity is taken into account in an isotropic manner, i.e. indirectly, an N-dimensional problem becomes essentially reduced to a one-dimensional case. This method has unknown prospects for a description of real polyatomic molecules.

3 Polynomials of Quantum Numbers

The Principles of the Theory

Reasoning from the results obtained in Refs. [16–19], we revise the approximations used in the stationary perturbation theory. The anharmonicity of normal vibrations is considered to imply an expansion of a perturbation in a power series in terms of normal coordinates q_i. For a one-dimensional case, the Hamiltonian is represented in a form

$$H = H^0 + \hbar\omega \sum_{p>0} \lambda^p a_p \xi^{p+2},$$

in which H^0 is the harmonic Hamiltonian with eigenvalues E_n^0 and eigenfunctions $|n\rangle$, ω is the frequency of harmonic vibrations, $\xi = \sqrt{2}q$ is a convenient vibrational variable, λ is a small parameter characterizing the perturbation order and a_p are dimensionless anharmonic force parameters including factor $2^{-(p+2)/2}$. The anharmonic Hamiltonian for an r-dimensional case is analogously written as

$$H = H^0 + \sum_{p>0} \lambda^p \sum_{(j_1 j_2 \cdots j_r)p+2} a_{j_1 j_2 \cdots j_r} \xi_1^{j_1} \xi_2^{j_2} \cdots \xi_r^{j_r}.$$

On denoting all vibrational variables as ξ, the Hamiltonian can be written in a general form

$$H = H^0 + \sum_{p>0} G_p(\xi)\lambda^p.$$

The equation for eigenvalues $E_n(\lambda)$ and eigenfunctions $|n,\lambda)$ of Hamiltonian H is differentiated with respect to parameter λ:

$$\sum_p pG_p(\xi)\lambda^{p-1}|n, \lambda) - \frac{\partial E_n(\lambda)}{\partial \lambda}|n, \lambda) = (E_n(\lambda) - H)\frac{\partial}{\partial \lambda}|n, \lambda).$$

Vector $|n,\lambda)$, which is terminated with a parenthesis, characterizes the exact state. For an infinitesimal change in parameter λ, we obtain vector $|n,\lambda + \delta\lambda)$, which is represented as a series expansion $\sum_m A_{mn}(\delta\lambda)|m, \lambda)$, i.e.

$$|n, \lambda + \delta\lambda) = A_{nn}(\delta\lambda)|n, \lambda) + \sum_{m \neq n} A_{mn}(\delta\lambda)|m, \lambda).$$

Quantum Theory of Anharmonic Effects in Molecules. DOI: http://dx.doi.org/10.1016/B978-0-12-397912-4.00003-9

This expansion is universally valid by virtue of the completeness of the eigenfunctions. Coefficients A_{mn} are related through an expression for normalization

$$|A_{nn}|^2 + \sum_{m\neq n} |A_{mn}|^2 = 1.$$

When $\delta\lambda = 0$, we have $A_{nn} = 1$ and $A_{mn} = 0$; hence, there exists a non-zero limit for the ratio $A_{mn}/\delta\lambda$ as $\delta\lambda \to 0$, which is by definition equal to C_{mn}. As a result, we obtain $A_{nn} = 1 - \text{const.} \cdot \delta\lambda^2 + \cdots$ and

$$\frac{\partial}{\partial\lambda}|n,\lambda) = \lim_{\delta\lambda\to 0} \frac{|n,\lambda+\delta\lambda) - |n,\lambda)}{\delta\lambda} = \sum_{m\neq n} C_{mn}|m,\lambda).$$

Returning to the differentiated equation for the eigenvalues and eigenfunctions, we have

$$\sum_p pG_p(\xi)\lambda^{p-1}|n,\lambda) - \frac{\partial E_n(\lambda)}{\partial\lambda}|n,\lambda) = \sum_{m\neq n} C_{mn}(E_n(\lambda) - E_m(\lambda))|m,\lambda).$$

Using this equation, we find

$$C_{mn} = \sum_p p\lambda^{p-1} \frac{(m,\lambda|G_p(\xi)|n,\lambda)}{E_n(\lambda) - E_m(\lambda)},$$

and determine the exact expansions for $\partial|n,\lambda)/\partial\lambda$ and $\partial E_n(\lambda)/\partial\lambda$ with respect to parameter λ. As a consequence, the required approximations of perturbation theory for E_n^α and $|n,\alpha\rangle$, which were introduced through these series expansions,

$$E_n(\lambda) = E_n^0 + \sum_{\alpha>0} \lambda^\alpha E_n^\alpha \quad \text{and} \quad |n,\lambda) = |n\rangle + \sum_{\alpha>0} \lambda^\alpha |n,\alpha\rangle,$$

take the form

$$\begin{aligned} E_n^\alpha &= \frac{1}{\alpha} \sum_{(p\beta\gamma)\alpha} p\langle n,\beta|G_p(\xi)|n,\gamma\rangle, \\ |n,\alpha\rangle &= \frac{1}{\alpha} \sum_{(pq\beta\gamma\nu)\alpha} \sum_{m\neq n} p\Delta_q(n,m)\langle m,\beta|G_p(\xi)|n,\gamma\rangle|m,\nu\rangle, \end{aligned} \tag{3.1}$$

in which

$$\begin{aligned} \Delta_q(n,m) &= \frac{1}{q!}\frac{\partial^q}{\partial\lambda^q}\left[\frac{1}{E_n(\lambda) - E_m(\lambda)}\right]_{\lambda=0} \\ &= \frac{1}{E_n^0 - E_m^0} \sum_i \sum_{(\alpha_1\alpha_2\cdots\alpha_i)q} \prod_{e=1}^{i} \frac{E_m^{\alpha_e} - E_n^{\alpha_e}}{E_n^0 - E_m^0}, \quad \Delta_0(n,m) = \frac{1}{E_n^0 - E_m^0}. \end{aligned}$$

Here, $\alpha > 0$ and the relation for $\Delta_q(n,m)$ is based on a simple expansion $(1-x)^{-1} = 1 + x + x^2 + \cdots$, which holds for $|x| < 1$.

For a one-dimensional case, we have $G_p = \hbar\omega a_p \xi^{p+2}$; according to Eq. (3.1), the first-order corrections to the vibrational energy and the corresponding function are determined by the matrix elements of quantity $\hbar\omega a_1 \xi^3$. Clearly, $E_n^1 = 0$ and

$$|n,1\rangle = a_1\left(\frac{1}{3}\sqrt{n(n-1)(n-2)}|n-3\rangle + 3n^{3/2}|n-1\rangle - 3(n+1)^{3/2}|n+1\rangle\right.$$
$$\left.-\frac{1}{3}\sqrt{(n+1)(n+2)(n+3)}|n+3\rangle\right), \qquad (3.2)$$

in which $|n\rangle$ is the state vector of a harmonic oscillator. For the r-dimensional case, one obtains the result in first order with a complicated perturbation function

$$\hbar\sum_i \omega_i a_i \xi_i^3 + \hbar\sum_{ij} a_{ij}\xi_i^2\xi_j + \hbar\sum_{ijk} a_{ijk}\xi_i\xi_j\xi_k,$$

in which ω_i denote the harmonic frequencies; a_i, a_{ij} and a_{ijk} are force parameters of which $a_{ii} = 0$ and $a_{ijk} = 0$ for $i \geq j \geq k$. Clearly, $E_n^1 = 0$ and

$$|n,1\rangle = \sum_i |n_1\rangle\cdots|n_i,1\rangle\cdots|n_r\rangle + |n,\mathrm{A}\rangle + |n,\mathrm{B}\rangle + |n,\mathrm{C}\rangle, \qquad (3.3)$$

in which n implies all quantum numbers from n_1 to n_r. Here, $|n_i,1\rangle$ is determined by quantity $\hbar\omega_i a_i \xi_i^3$ and is exactly equal to Eq. (3.2). Corrections $|n,\mathrm{A}\rangle$ and $|n,\mathrm{B}\rangle$ appear through $\hbar\sum_{ij} a_{ij}\xi_i^2\xi_j$; as a result,

$$|n,\mathrm{A}\rangle = \sum_{ij}\frac{2a_{ij}}{\omega_j}\left(n_i + \frac{1}{2}\right)(\eta_j - \eta_j^+)|n\rangle,$$
$$|n,\mathrm{B}\rangle = \sum_{ij} a_{ij}\left(\frac{\eta_i^2\eta_j - (\eta_i^+)^2\eta_j^+}{2\omega_i + \omega_j} + \frac{\eta_i^2\eta_j^+ - (\eta_i^+)^2\eta_j}{2\omega_i - \omega_j}\right)|n\rangle,$$

in which appear operators η_i^+ for creation and η_i for destruction:

$$\eta_i^+|n\rangle = \sqrt{n_i + 1}|n_1, n_2, \ldots, n_i + 1, \ldots, n_r\rangle$$

and

$$\eta_i|n\rangle = \sqrt{n_i}|n_1, n_2, \ldots, n_i - 1, \ldots, n_r\rangle.$$

Taking into account $\hbar \sum_{ijk} a_{ijk}\xi_i\xi_j\xi_k$, we find

$$|n,\mathrm{C}\rangle = \sum_{ijk} a_{ijk}\left(\frac{\eta_i\eta_j\eta_k - \eta_i^+\eta_j^+\eta_k^+}{\omega_i+\omega_j+\omega_k} + \frac{\eta_i\eta_j\eta_k^+ - \eta_i^+\eta_j^+\eta_k}{\omega_i+\omega_j-\omega_k} + \frac{\eta_i\eta_j^+\eta_k - \eta_i^+\eta_j\eta_k^+}{\omega_i-\omega_j+\omega_k} + \frac{\eta_i^+\eta_j\eta_k - \eta_i\eta_j^+\eta_k^+}{-\omega_i+\omega_j+\omega_k}\right)|n\rangle.$$

Such a construction of the first correction to the wave function becomes convenient and might readily have an interpretation: namely, for normal vibration i, $|n_i,1\rangle$ characterizes its own anharmonicity. Quantity $|n,\mathrm{A}\rangle$ represents an expansion in terms of the states, in each of which only a single normal vibration becomes perturbed. Vectors $|n,\mathrm{B}\rangle$ and $|n,\mathrm{C}\rangle$ are expansions in states with two and three perturbed vibrations, respectively.

Recurrence Equations

We specify the parity of the harmonic state vector so that $|n \pm k\rangle$ at a fixed quantum number n has the parity of number k; for instance, $|n-1\rangle$ and $|n+3\rangle$ characterize odd states, $|n-2\rangle$ and $|n\rangle$ even states. Moreover, for $m \geq n$, we introduce factor g_{nm} as

$$g_{nm} = (n+1)(n+2)\cdots(m-1)m, \quad g_{nn} = 1.$$

The system of equations (3.1) for the case of one variable, when $\alpha > 0$ and quantity $\Delta_q(n,m)$ is chosen to be dimensionless, takes the following form:

$$\begin{aligned} E_n^\alpha &= \frac{\hbar\omega}{\alpha}\sum_{(p\beta\gamma)\alpha} pa_p\langle n,\beta|\xi^{p+2}|n,\gamma\rangle, \\ |n,\alpha\rangle &= \frac{1}{\alpha}\sum_{(pq\beta\gamma\nu)\alpha}\sum_{m\neq n} pa_p\Delta_q(n,m)\langle m,\beta|\xi^{p+2}|n,\gamma\rangle|m,\nu\rangle, \\ \Delta_q(n,m) &= \hbar\omega\sum_i\sum_{(\alpha_1\cdots\alpha_i)q}\frac{(E_m^{\alpha_1}-E_n^{\alpha_1})\cdots(E_m^{\alpha_i}-E_n^{\alpha_i})}{(E_n^0-E_m^0)^{i+1}}, \quad \Delta_0(n,m) = \frac{\hbar\omega}{E_n^0-E_m^0}. \end{aligned} \tag{3.4}$$

In particular, it follows that $E_n^1 = 0$ and $|n,1\rangle$ contains states of only odd parity (see Eq. (3.2)). Neglecting a small constant, for the second-order correction, we have

$$E_n^2 = -\hbar\omega(30a_1^2 - 6a_2)\left(n+\frac{1}{2}\right)^2;$$

and states of only even parity are involved in function $|n,2\rangle$.

In the general case, vector $|n,\alpha\rangle$ involves states $|n \pm k\rangle$ of parity α with $k = 3\alpha$ as a bound of the expansion. The necessary condition for this generalization is identity $\Delta_{2q-1} = 0$, which follows from trivial equality $E_n^{2\alpha-1} = 0$. To prove this assertion, it suffices to consider Eq. (3.4) more thoroughly. Under an assumption that function $|n,\gamma\rangle$ is expanded in terms of states $|n \pm c_\gamma\rangle$, in which number c_γ has parity γ, we readily obtain condition $m \pm c_\beta = n \pm c_\gamma \pm c_p$, which must be satisfied by the non-zero matrix elements $\langle m,\beta|\xi^{p+2}|n,\gamma\rangle$. Function $|n,\alpha\rangle$ is therefore an expansion in terms of vectors $|n \pm c_p \pm c_\beta \pm c_\gamma \pm c_\nu\rangle$, whereas function $|n,\alpha\rangle$ is formed by states $|n \pm c_\alpha\rangle$. Consequently, we have an equality $\pm c_\alpha = \pm c_p \pm c_\beta \pm c_\gamma \pm c_\nu$. Only for even values of q this result does not contradict a relation $\alpha = p + q + \beta + \gamma + \nu$. Quantity Δ_{2q-1} is thus identically equal to zero. The converse statement is also obviously true: if correction $|n,\alpha\rangle$ contains states $|n \pm c_\alpha\rangle$, we have $E_n^{2\alpha-1} = 0$ for an arbitrary odd correction. This conclusion establishes the validity of the above assertion.

To demonstrate the above analysis, we rewrite $|n,\alpha\rangle$ in accordance with Eq. (3.4) in this form,

$$|n,\alpha\rangle = a_1^\alpha \sum_{m_1 m_2 \cdots m_\alpha} h(n,m,\alpha)\langle n|\xi^3|m_1\rangle\langle m_1|\xi^3|m_2\rangle \cdots \langle m_{\alpha-1}|\xi^3|m_\alpha\rangle|m_\alpha\rangle + \cdots,$$

in which an explicit form of quantity $h(n,m,\alpha)$ has no special interest. We see that vectors $|n \pm 3\alpha\rangle$ are really bound states in an expansion of $|n,\alpha\rangle$ in harmonic state vectors. Moreover, in calculating the amplitude of a harmonic state, e.g. state $|n+k\rangle$ of function $|n,\alpha\rangle$, it is necessary to sum various products

$$\langle n|\xi^s|\ell\rangle\langle\ell|\xi^q|r\rangle \cdots \langle u|\xi^p|n+k\rangle,$$

which are proportional to $\sqrt{g_{n,n+k}}$ through a relation $\langle\ell|\xi^q|\ell+k\rangle \sim \sqrt{g_{\ell,\ell+k}}$. To introduce polynomials $\Pi^s_{\alpha\beta}(n,m)$ at $m \geq n$ is thus convenient in the following form:

$$\langle n,\alpha|\xi^s|m,\beta\rangle = \sqrt{g_{nm}}\Pi^s_{\alpha\beta}(n,m). \tag{3.5}$$

In this definition, one must distinguish between the orders of indices α and β, and between the orders of numbers n and m.

To convert from matrix elements to polynomials, it suffices to multiply the left side of relation (3.4) for $|n,\alpha\rangle$ by expression $\langle\ell,\mu|\xi^s$, i.e.

$$\langle\ell,\mu|\xi^s|n,\alpha\rangle = \frac{1}{\alpha}\sum_{(pq\beta\gamma\nu)\alpha}\sum_{m\neq n} pa_p\Delta_q(n,m)\langle m,\beta|\xi^{p+2}|n,\gamma\rangle\langle\ell,\mu|\xi^s|m,\nu\rangle.$$

On substituting Eq. (3.5) and making elementary algebraic transformations, we obtain the required recurrence equations,

$$\Pi^{s}_{\mu\alpha}(\ell,n)=\frac{1}{\alpha}\sum_{(pq\beta\gamma\nu)\alpha} pa_p\left[\sum_{m<\ell} g_{m\ell}\Delta_q\Pi^{s}_{\nu\mu}(m,\ell)\Pi^{p+2}_{\beta\gamma}(m,n)\right.$$
$$\left.+\sum_{\ell\le m<n}\Delta_q\Pi^{s}_{\mu\nu}(\ell,m)\Pi^{p+2}_{\beta\gamma}(m,n)+\sum_{m>n} g_{nm}\Delta_q\Pi^{s}_{\mu\nu}(\ell,m)\Pi^{p+2}_{\gamma\beta}(n,m)\right],$$

$$\Pi^{s}_{\alpha\mu}(n,\ell)=\frac{1}{\alpha}\sum_{(pq\beta\gamma\nu)\alpha} pa_p\left[\sum_{m<n} g_{mn}\Delta_q\Pi^{s}_{\nu\mu}(m,\ell)\Pi^{p+2}_{\beta\gamma}(m,n)\right.$$
$$\left.+\sum_{n<m\le\ell}\Delta_q\Pi^{s}_{\nu\mu}(m,\ell)\Pi^{p+2}_{\gamma\beta}(n,m)+\sum_{m>\ell} g_{\ell m}\Delta_q\Pi^{s}_{\mu\nu}(\ell,m)\Pi^{p+2}_{\gamma\beta}(n,m)\right].$$

To determine the polynomials, we set $\ell=n-k$ in the first equation and $\ell=n+k$ in the second equation; in this case, the polynomials clearly become expansions in n to various powers.

According to definition (3.5), some polynomials equal zero; namely, function $|n\pm k,\beta\rangle$ is expanded in terms of states $|n\pm k\pm c_\beta\rangle$, in which number c_β has parity β. Correspondingly, quantity $\xi^s|n\pm k,\beta\rangle$ is an expansion in terms of harmonic vectors $|n\pm k\pm c_\beta\pm c_s\rangle$. Matrix elements $\langle n,\alpha|\xi^s|n\pm k,\beta\rangle$ are defined by quantities $\langle n\pm c_\alpha|n\pm k\pm c_\beta\pm c_s\rangle$, yielding $k=\pm c_s\pm c_\alpha\pm c_\beta$. In polynomials $\Pi^{s}_{\beta\alpha}(n-k,n)$ and $\Pi^{s}_{\alpha\beta}(n,n+k)$, number k has therefore the parity of number $s+\alpha+\beta$. The maximum or bounding value of k is also determined by numbers s, α, β and equals $3(\alpha+\beta)+s$. Polynomials $\Pi^{s}_{\beta\alpha}(n-k,n)$ and $\Pi^{s}_{\alpha\beta}(n,n+k)$ for $k>3(\alpha+\beta)+s$ are hence identically equal to zero.

Apart from a direct calculation of the polynomials with the use of these recurrence equations, these polynomials are expressible through additional relations. First, as vectors $|n)$ represent an orthogonal normalized system, we write

$$(n|m)=\sum_{\beta\gamma}\langle n,\beta|m,\gamma\rangle=\delta_{nm};$$

hence,

$$\sum_{(\beta\gamma)\alpha}\Pi_{\beta\gamma}(n,m)=0,\quad \alpha>0, \tag{3.6}$$

in which $\Pi_{\alpha\beta}(n,m)=\Pi^{0}_{\alpha\beta}(n,m)$. Identity (3.6) provides the normalization of the wave function in perturbation theory of any order.

Second, through a condition $\sum_\ell|\ell\rangle\langle\ell|=1$, we have

$$\langle n,\alpha|\xi^{s+q}|m,\beta\rangle=\sum_{\ell}\langle n,\alpha|\xi^{s}|\ell\rangle\langle\ell|\xi^{q}|m,\beta\rangle.$$

Then, on converting to polynomials, we obtain this addition theorem:

$$\Pi^{s+q}_{\alpha\beta}(n,m) = \sum_{\ell<n} g_{\ell n}\Pi^{s}_{0\alpha}(\ell,n)\Pi^{q}_{0\beta}(\ell,m) + \sum_{n\le\ell\le m} \Pi^{s}_{\alpha 0}(n,\ell)\Pi^{q}_{0\beta}(\ell,m)$$
$$+ \sum_{\ell>m} g_{m\ell}\Pi^{s}_{\alpha 0}(n,\ell)\Pi^{q}_{\beta 0}(m,\ell).$$

Within the formalism under consideration, an important role is evidently played, as it must be, by polynomials $\Pi_{0\alpha}(n-k,n)$ and $\Pi_{\alpha 0}(n,n+k)$, and, of course, harmonic polynomials $\Pi^{s}(n,m) = \Pi^{s}_{00}(n,m)$.

We proceed to formulate the principal definitions in terms of the polynomial language. It is convenient to suppose that $\lambda = 1$; in this case, coefficients a_p pertain to a small order, i.e. $a_p \sim \lambda^p$. Furthermore, only in exceptional cases we show parameter λ explicitly. The energy of anharmonic vibrations in the one-dimensional case can be written as

$$E_n = \hbar\omega\left(n+\frac{1}{2}\right) + \sum_{\alpha} E^{\alpha}_{n}, \quad E^{\alpha}_{n} = \frac{\hbar\omega}{\alpha}\sum_{(p\beta\gamma)\alpha} p a_p \Pi^{p+2}_{\beta\gamma}(n,n).$$

The summation is clearly taken solely with respect to even values of α because only even corrections to the vibrational energy are non-zero. This circumstance is important. The first-order correction to the energy is equal to zero; the second-order correction is proportional to a_1^2 and a_2, and the next non-vanishing correction is linear in a_1^4, $a_1^2 a_2$, a_2^2, $a_1 a_3$ and a_4, i.e. the correction proportional to a_1^3, $a_1 a_2$ and a_3 vanishes. Continuing in this manner, we obtain the exact relations for corrections of higher order in complete agreement with the experimental data.

The arbitrary correction to the wave function takes a simple form,

$$|n,\alpha\rangle = \sum_{k=0}^{3\alpha} \sqrt{g_{n-k,n}}\,\Pi_{0\alpha}(n-k,n)|n-k\rangle + \sum_{k=1}^{3\alpha} \sqrt{g_{n,n+k}}\,\Pi_{\alpha 0}(n,n+k)|n+k\rangle,$$

in which the summation is extended over values of k with a parity identical to that of α. Despite the apparent triviality of this expression for $|n,\alpha\rangle$, the wave function has a latent role in the polynomial formalism.

We consider finally a function $f(\xi)$, which is expanded as a power series in ξ:

$$f(\xi) = \sum_{s} \frac{2^{-s/2}}{s!} f^{(s)}\xi^{s}.$$

Derivatives $f^{(s)}$ are chosen so that this expansion is a Taylor series for variable q. On setting $F_s = 2^{-s/2} f^{(s)}/s!$, the matrix element can be expressed as

$$(n|f|n+k) = \sum_{s\beta\gamma} F_s\langle n,\beta|\xi^{s}|n+k,\gamma\rangle.$$

From this formula and definition (3.5), we obtain

$$(n|f|n+k) = \sqrt{g_{n,n+k}} \sum_{s\alpha} \sum_{(\beta\gamma)\alpha} F_s \Pi^s_{\beta\gamma}(n, n+k). \tag{3.7}$$

This general scheme to construct the new formalism is sufficiently simple. The recurrence equations thus derived enable a definition of an arbitrary polynomial in an explicit form. By this means, we calculate all desired polynomials; the sought expressions, in particular, for the eigenvalues and eigenfunctions, then depend only on the accuracy of the required approximation [19].

Many-Dimensional Case

The calculations in perturbation theory for a system with variables of arbitrary number differ substantially from those for a one-dimensional case even in the first order. The calculation of the matrix elements, for instance, with the help of functions (3.3), is accompanied by competition among various mechanical approximations and yields cumbersome expressions. Furthermore, the derivatives of the dipolar moment that have a maximum influence on the matrix elements are unknown. Using the polynomial technique, we consider these questions in detail.

Returning to Eq. (3.1), we consider an arbitrary correction $|n,\alpha\rangle$ to the function in the harmonic approximation. Vector $|n,\alpha\rangle$ is constructed from all possible states $|n_1 \pm \ell_1,\ldots,n_i \pm \ell_i,\ldots,n_r \pm \ell_r\rangle$, in which $\ell_1 + \ell_2 + \cdots + \ell_r \leq 3\alpha$. Functions $|n_i + \ell_i\rangle$ and $|n_i - \ell_i\rangle$ are here multiplied by factors $\sqrt{g_{n_i,n_i+\ell_i}}$ and $\sqrt{g_{n_i-\ell_i,n_i}}$, respectively. Recall that state vector $|n_1,n_2,\ldots,n_r\rangle$ is a product of individual functions $|n_1\rangle, |n_2\rangle,\ldots,|n_r\rangle$. We construct an arbitrary matrix element between separate corrections $|n,\alpha\rangle$ and $|n+k,\beta\rangle$:

$$\langle n,\alpha|\xi_1^{s_1}\xi_2^{s_2}\cdots\xi_r^{s_r}|n+k,\beta\rangle.$$

This matrix element is determined essentially by elements $\langle n_i \pm \ell_i|\xi_i^{s_i}|n_i \pm p_i + k_i\rangle$:

$$\sqrt{g_{n_i,n_i+\ell_i} g_{n_i,n_i+p_i+k_i}}\langle n_i + \ell_i|\xi_i^{s_i}|n_i + p_i + k_i\rangle,$$

$$\sqrt{g_{n_i-\ell_i,n_i} g_{n_i,n_i+p_i+k_i}}\langle n_i - \ell_i|\xi_i^{s_i}|n_i + p_i + k_i\rangle, \text{ etc.}$$

According to a relation $\langle \ell|\xi^s|\ell+p\rangle \sim \sqrt{g_{\ell,\ell+p}}$, these matrix elements are proportional to a factor $\sqrt{g_{n_i,n_i+k_i}}$; hence,

$$\langle n,\alpha|\xi_1^{s_1}\xi_2^{s_2}\cdots\xi_r^{s_r}|n+k,\beta\rangle = \sqrt{g_{n,n+k}}\,\Pi^s_{\alpha\beta}(n,n+k),$$

in which the polynomial

$$\Pi^{s_1 s_2 \cdots s_r}_{\alpha\beta}(n_1, n_1+k_1; n_2, n_2+k_2; \ldots; n_r, n_r+k_r)$$

of quantum numbers n_1, n_2,. . ., n_r is designated as $\Pi^s_{\alpha\beta}(n, n+k)$. Moreover, here, $g_{n,n+k} \equiv g_{n_1,n_1+k_1} g_{n_2,n_2+k_2} \cdots g_{n_r,n_r+k_r}$ and sets $\{s_1, s_2, \ldots, s_r\}$ and $\{k_1, k_2, \ldots, k_r\}$ are denoted by s and k, respectively.

This definition of polynomials produces a general selection rule for k_i. Matrix element $\langle n, \alpha | \xi_1^{s_1} \xi_2^{s_2} \cdots \xi_r^{s_r} | n+k, \beta \rangle$ is determined by a sum of matrix elements $\langle n_i \pm \ell_i | n_i \pm p_i \pm c_{si} + k_i \rangle$ in various combinations, in which c_{si} has the parity of number s_i; equality $k_i = \pm \ell_i \pm p_i \pm c_{si}$ hence becomes satisfied. On summing this equality with respect to i, we obtain the desired rule,

$$\sum_i k_i \leftrightarrow \alpha + \beta + \sum_i s_i,$$

which asserts that polynomials $\Pi^s_{\beta\alpha}(n-k, n)$ and $\Pi^s_{\alpha\beta}(n, n+k)$ differ from zero only for a case in which $\sum_i k_i$ has the parity of number $\alpha + \beta + \sum_i s_i$.

By analogy with the one-dimensional case, from system (3.1), we convert from the matrix elements to polynomials and derive the corresponding recurrence equations,

$$\Pi^s_{\mu\alpha}(\ell, n) = \frac{1}{\alpha} \sum_{(pq\beta\gamma\nu)\alpha} p \sum_{(j)p+2} a_j \left[\sum_{m<\ell} g_{m\ell} \Delta_q \Pi^s_{\nu\mu}(m, \ell) \Pi^j_{\beta\gamma}(m, n) \right.$$
$$\left. + \sum_{\ell \le m < n} \Delta_q \Pi^s_{\mu\nu}(\ell, m) \Pi^j_{\beta\gamma}(m, n) + \sum_{m>n} g_{nm} \Delta_q \Pi^s_{\mu\nu}(\ell, m) \Pi^j_{\gamma\beta}(n, m) \right],$$

$$\Pi^s_{\alpha\mu}(n, \ell) = \frac{1}{\alpha} \sum_{(pq\beta\gamma\nu)\alpha} p \sum_{(j)p+2} a_j \left[\sum_{m<n} g_{mn} \Delta_q \Pi^s_{\nu\mu}(m, \ell) \Pi^j_{\beta\gamma}(m, n) \right.$$
$$\left. + \sum_{n<m\le\ell} \Delta_q \Pi^s_{\nu\mu}(m, \ell) \Pi^j_{\gamma\beta}(n, m) + \sum_{m>\ell} g_{\ell m} \Delta_q \Pi^s_{\mu\nu}(\ell, m) \Pi^j_{\gamma\beta}(n, m) \right],$$

in which j implies a set $\{j_1, j_2, \ldots, j_r\}$, and factor $\Delta_q(n,m)$ is given by the expression

$$\sum_i \sum_{(\alpha_1\alpha_2\cdots\alpha_i)q} \frac{(E_m^{\alpha_1} - E_n^{\alpha_1})(E_m^{\alpha_2} - E_n^{\alpha_2}) \cdots (E_m^{\alpha_i} - E_n^{\alpha_i})}{(E_n^0 - E_m^0)^{i+1}},$$

with $\Delta_0(n,m) = (E_n^0 - E_m^0)^{-1}$. Quantity E_n^α has a linear dependence on $\Pi^j_{\beta\gamma}(n, n)$ and represents the correction of order α to the energy of harmonic vibrations E_n^0.

The recurrence equations completely retain their form; the only difference is that term $\hbar\omega a_p \Pi^{p+2}$ transforms into term $\sum a_{j_1 j_2 \cdots j_r} \Pi^{j_1 j_2 \cdots j_r}$. Setting $\ell = n - k$, i.e. $\ell_1 = n_1 - k_1$, $\ell_2 = n_2 - k_2$, etc., in the first equation and $\ell = n + k$ in the second one, we obtain the expansions of the polynomials in a power series in quantum numbers n_1, n_2,. . ., n_r.

Having applied the above correspondence, we write the exact energy of anharmonic vibrations as

$$E_n = \hbar \sum_{i=1}^{r} \omega_i \left(n_i + \frac{1}{2} \right) + \sum_{\alpha} E_n^{\alpha}, \quad E_n^{\alpha} = \frac{1}{\alpha} \sum_{(p\beta\gamma)\alpha} p \sum_{(j)p+2} a_j \Pi_{\beta\gamma}^{j}(n,n).$$

In terms of polynomials, the arbitrary correction to wave function $|n,\alpha\rangle$, the polynomial addition theorem and identity (3.6) that imposes normalization of the wave function remain valid and, moreover, exactly retain their form. In the final expressions $\lambda = 1$.

We 'translate' the formula for matrix elements of an arbitrary coordinate function

$$f = \sum_{\ell} \sum_{(s)\ell} \frac{2^{-\ell/2}}{\ell!} f_{s_1 s_2 \cdots s_r}^{(\ell)} \xi_1^{s_1} \xi_2^{s_2} \cdots \xi_r^{s_r}.$$

Here, $f_s^{(\ell)}$ are ordinary derivatives in a Taylor-series expansion of function f in normal coordinates q_i. If $|n) = \sum_{\alpha} |n,\alpha\rangle$, in which $|n,0\rangle$ is the harmonic state vector, then

$$(n|f|n+k) = \sum_{\ell} \sum_{(s)\ell} \frac{2^{-\ell/2}}{\ell!} f_{s_1 s_2 \cdots s_r}^{(\ell)} \sum_{\alpha\beta} \langle n,\alpha | \xi_1^{s_1} \xi_2^{s_2} \cdots \xi_r^{s_r} | n+k,\beta \rangle.$$

Having used the polynomial definition (3.5), we derive the formula

$$(n|f|n+k) = \sqrt{g_{n,n+k}} \sum_{\ell} \sum_{(s)\ell} \frac{2^{-\ell/2}}{\ell!} f_s^{(\ell)} \sum_{\alpha\beta} \Pi_{\alpha\beta}^{s}(n,n+k),$$

which has an obvious coincidence with Eq. (3.7).

The Problem of Degenerate States

With regard to degenerate states in the polynomial formalism, we consider whether the perturbation theory developed in Ref. [16] retains its advantages over the conventional theory. We begin with the general case [18] for which the exact equation for eigenvalues $E_{nx}(\lambda)$ and eigenfunctions $|nx,\lambda)$ has the form

$$\left(H^0 + \sum_{p>0} \lambda^p G_p \right) |nx,\lambda) = E_{nx}(\lambda) |nx,\lambda).$$

Here, G_p are perturbations of various orders in small parameter λ, and the zero-order Hamiltonian H^0 has degenerate eigenvalues $E_n^0 (= E_{nx}^0)$ with functions $|nx\rangle$, in which index x numbers the degenerate states corresponding to level n. Repeating, to some extent, the reasoning used in our derivation of the basic relations of the perturbation theory, we construct the algebraic solutions for $E_{nx}(\lambda)$ and $|nx,\lambda)$ in the form of series in λ to various powers with the inclusion of degeneracy.

The method is formally simple. Differentiating the exact eigenvalue equation with respect to λ and using an expansion in the exact eigenvectors,

$$\frac{\partial}{\partial\lambda}|nx,\lambda) = \sum_{my\neq nx} C_{my,nx}|my,\lambda),$$

we generate the following system:

$$\begin{aligned}\frac{\partial E_{nx}(\lambda)}{\partial\lambda} &= \sum_p p\lambda^{p-1}(nx,\lambda|G_p|nx,\lambda),\\ \frac{\partial}{\partial\lambda}|nx,\lambda) &= \sum_{my\neq nx}\sum_p p\lambda^{p-1}\frac{(my,\lambda|G_p|nx,\lambda)}{E_{nx}(\lambda)-E_{my}(\lambda)}|my,\lambda).\end{aligned}$$

Assuming, as is customary, that

$$E_{nx}(\lambda) = E_n^0 + \sum_{\alpha>0}\lambda^\alpha E_{nx}^\alpha \text{ and } |nx,\lambda) = |nx\rangle + \sum_{\alpha>0}\lambda^\alpha|nx,\alpha\rangle,$$

we obtain the desired expressions for corrections E_{nx}^α and $|nx,\alpha\rangle$ for the degenerate case in a recurrent manner,

$$\begin{aligned}E_{nx}^\alpha &= \frac{1}{\alpha}\sum_{(p\beta\gamma)\alpha} p\langle nx,\beta|G_p|nx,\gamma\rangle,\\ |nx,\alpha\rangle &= \frac{1}{\alpha}\sum_{y;m\neq n}\sum_{(pq\beta\gamma\nu)\alpha} p\Delta_q^0(nx,my)\langle my,\beta|G_p|nx,\gamma\rangle|my,\nu\rangle\\ &\quad+\frac{1}{\alpha}\sum_{y\neq x}\sum_{(pq\beta\gamma\nu)\alpha>\sigma} p\Delta_q^\sigma(nx,ny)\langle ny,\beta|G_p|nx,\gamma\rangle|ny,\nu\rangle,\end{aligned} \tag{3.8}$$

in which $\Delta_0^\sigma(nx,my) = (E_{nx}^\sigma - E_{my}^\sigma)^{-1}$ and, for $q>0$,

$$\Delta_q^\sigma(nx,my) = \frac{1}{E_{nx}^\sigma - E_{my}^\sigma}\sum_i\sum_{(\alpha_1\alpha_2\cdots\alpha_i)q}\prod_{e=1}^{i}\frac{E_{my}^{\sigma+\alpha_e}-E_{nx}^{\sigma+\alpha_e}}{E_{nx}^\sigma - E_{my}^\sigma}.$$

Factor $\Delta_q^0(nx,my)$ retains its preceding meaning,

$$\frac{1}{E_{nx}(\lambda)-E_{my}(\lambda)} = \frac{1}{(E_n^0-E_m^0)+(E_{nx}^1-E_{my}^1)\lambda+\cdots} = \sum_q \lambda^q\Delta_q^0(nx,my),$$

whereas for the degenerate states of level n,

$$\frac{1}{E_{nx}(\lambda)-E_{ny}(\lambda)}=\frac{1}{(E^{\sigma}_{nx}-E^{\sigma}_{ny})\lambda^{\sigma}+(E^{\sigma+1}_{nx}-E^{\sigma+1}_{ny})\lambda^{\sigma+1}+\cdots}=\sum_{q}\lambda^{q-\sigma}\Delta^{\sigma}_{q}(nx,ny),$$

in which $\sigma\geq 1$. If, for instance, $E^{1}_{nx}-E^{1}_{ny}\neq 0$ for all $y\neq x$, then $\sigma=1$. If $E^{1}_{nx}-E^{1}_{ny}=0$, one must consider $\sigma=2$, etc.

Our theory thus remains valid in the presence of degenerate states. Because the expansion coefficients of $\partial|nx,\lambda\rangle/\partial\lambda$ in terms of $|my,\lambda\rangle$ contain only differences $E_{nx}(\lambda)-E_{my}(\lambda)$ as the denominators, the developed method possesses a further merit: it allows one to eliminate zeros in the denominators upon summing over the degenerate states. Comparison of the results obtained here with the non-degenerate case [16] shows a formal similarity of the expressions for E^{α}_{nx} and an evident distinction in calculating the corrections $|nx,\alpha\rangle$ involving an additional summation over the group of degenerate states beginning with $\alpha=\sigma+1$.

To elucidate the meaning of σ, we consider the first two corrections in detail; for the first correction,

$$|nx,1\rangle=\sum_{y;m\neq n}\frac{\langle my|G_1|nx\rangle}{E^{0}_{n}-E^{0}_{m}}|my\rangle+\left(\sum_{y\neq x}\frac{\langle ny|G_1|nx\rangle}{E^{0}_{nx}-E^{0}_{ny}}|ny\rangle=0\right).$$

The absence of terms with $\sigma=0$ implies that all matrix elements $\langle ny|G_1|nx\rangle$ vanish for $y\neq x$, which allows one to eliminate all zeros in denominators in the group of degenerate states. This result indicates that the correct functions of zero approximation $|nx\rangle$ were chosen as the basis functions. This operation is effected through an appropriate unitary transformation of the eigenfunctions on the basis of the solution of a secular equation [40]. Assuming that $\sigma=1$, we consider the second correction

$$|nx,2\rangle=\frac{1}{2}\sum_{y;m\neq n}\sum_{(pq\beta\gamma\nu)2}p\Delta^{0}_{q}(nx,my)\langle my,\beta|G_p|nx,\gamma\rangle|my,\nu\rangle$$
$$+\frac{1}{2}\sum_{y\neq x}\frac{\langle ny,1|G_1|nx\rangle+\langle ny|G_1|nx,1\rangle+2\langle ny|G_2|nx\rangle}{E^{1}_{nx}-E^{1}_{ny}}|ny\rangle.$$

If for some reason

$$E^{1}_{nx}-E^{1}_{ny}=\langle nx|G_1|nx\rangle-\langle ny|G_1|ny\rangle=0,$$

functions $|nx\rangle$ should be again unitarily transformed (with the aid of matrix U)

$$|nx'\rangle=\sum_{x}U_{xx'}|nx\rangle$$

so that

$$\langle ny', 1|G_1|nx'\rangle + \langle ny'|G_1|nx', 1\rangle + 2\langle ny'|G_2|nx'\rangle = 0.$$

We emphasize that, in the new basis set,

$$\langle ny'|G_1|nx'\rangle = \sum_{xy} U^*_{yy'} U_{xx'} \langle ny|G_1|nx\rangle = \sum_{x} U^*_{xy'} U_{xx'} \langle nx|G_1|nx\rangle = \langle nx|G_1|nx\rangle \delta_{y'x'},$$

i.e. all matrix elements $\langle ny'|G_1|nx'\rangle$ vanish as before, except the case $y' = x'$. Consequently, for $E^1_{nx} - E^1_{ny} = 0$, $\sigma = 2$ and the summation within the group of degenerate states should be performed only in calculating the third- and higher-order corrections $|nx, \alpha\rangle$ with $\alpha \geq 3$.

To eliminate trivial zeros in the denominators when $E^0_{nx} - E^0_{ny} = 0$, one must perform a unitary transformation of the basis set so that, for $y \neq x$,

$$\sum_{(p\beta\gamma)1} p\langle ny, \beta|G_p|nx, \gamma\rangle = \langle ny|G_1|nx\rangle = 0.$$

This basis set was used initially. If the degeneracy is removed, $\sigma = 1$; otherwise, when $E^1_{nx} - E^1_{ny} = 0$, the basis functions should again be transformed so that, in the new basis set,

$$\sum_{(p\beta\gamma)2} p\langle ny', \beta|G_p|nx', \gamma\rangle = 0,$$

consequently, $\sigma = 2$. This reasoning becomes generalized with a simple scheme ($\sigma = \sigma$):

$$E^{\sigma-1}_{nx} - E^{\sigma-1}_{ny} = 0 \rightarrow \sum_{(p\beta\gamma)\sigma} p\langle ny^{(\sigma-1)}, \beta|G_p|nx^{(\sigma-1)}, \gamma\rangle = 0.$$

After the value of σ is chosen on sequentially eliminating the zeros, the required corrections $|nx,\alpha\rangle$ and E^α_{nx} thus become reconstructed from the equations in system (3.8).

Introduction to a Theory of Anharmonicity

We apply the results obtained to the problem of anharmonicity of interest. In this case,

$$G_p = \sum_{(j_1 j_2 \cdots j_r)p+2} a_{j_1 j_2 \cdots j_r} \xi_1^{j_1} \xi_2^{j_2} \cdots \xi_r^{j_r},$$

the eigenvalues of H^0 are

$$E_n^0 = \hbar \sum_{i=1}^{r} \omega_i \left(n_i + \frac{1}{2} \right),$$

and the expansions in the harmonic-oscillator eigenvectors

$$|n_1, n_2, \ldots, n_r\rangle = |n_1\rangle |n_2\rangle \cdots |n_r\rangle$$

for which the coefficients of an appropriate unitary transformation $U^n_{xx_i}$

$$|nx\rangle = \sum_{x_1 x_2 \cdots x_r} U^n_{xx_i} |n_1 + x_1, \ldots, n_i + x_i, \ldots, n_r + x_r\rangle$$

should be chosen as the correct functions for initial states. Here, x_i are known integers (both positive and negative) that specify the complete set of degenerate vibrational states of level n and are determined from the conditions

$$\sum_{i=1}^{r} \omega_i x_i = 0.$$

We assume the vibrational levels to be degenerate; consequently, $\sigma > 0$. Then, according to Eq. (3.8),

$$|nx, 1\rangle = \sum_{y; m \neq n} \sum_{(j_1 \cdots j_r) 3} a_{j_1 j_2 \cdots j_r} \frac{\langle my | \xi_1^{j_1} \xi_2^{j_2} \cdots \xi_r^{j_r} | nx \rangle}{E_n^0 - E_m^0} |my\rangle.$$

Because $\sum_y (U^m_{yy_i}) * U^m_{yy_i'} = \delta_{y_i y_i'}$, correction $|nx,1\rangle$ is an expansion in these vectors:

$$\prod_{i=1}^{r} \sqrt{g_{n_i + x_i, n_i + x_i \pm \ell_i}} |n_i + x_i \pm \ell_i\rangle, \text{ in which } \ell_1 + \ell_2 + \cdots + \ell_r \leq 3.$$

Correction $|nx,2\rangle$ is represented with a similar expansion but with $\ell_1 + \ell_2 + \cdots + \ell_r \leq 6$. Finally, it is easily shown by induction that, in the general case, correction $|nx,\alpha\rangle$ is expanded in the functions

$$\prod_{i=1}^{r} \sqrt{g_{n_i + x_i, n_i + x_i \pm \ell_i}} |n_i + x_i \pm \ell_i\rangle, \text{ in which } \ell_1 + \ell_2 + \cdots + \ell_r \leq 3\alpha,$$

each correction contains coefficient $U^n_{xx_i}$ appearing necessarily upon summation over x_i.

We introduce this matrix element,

$$\mathrm{M}^{s_1 s_2 \cdots s_r}_{\alpha\beta}(nx, my) = \langle nx, \alpha | \xi_1^{s_1} \xi_2^{s_2} \cdots \xi_r^{s_r} | my, \beta \rangle.$$

In view of the properties of corrections $|nx,\alpha\rangle$ and $|my,\beta\rangle$ considered above, $M^s_{\alpha\beta}(nx,my)$ clearly comprises various elements $\langle n_i + x_i \pm \ell_i|\xi_i^{s_i}|m_i + y_i \pm \ell'_i\rangle$ with the corresponding factors $\sqrt{g_{n_i+x_i,n_i+x_i\,\pm\,\ell_i}}$ and $\sqrt{g_{m_i+y_i,m_i+y_i\,\pm\,\ell'_i}}$, but $\langle k|\xi^s|p\rangle \sim \sqrt{g_{kp}}$; consequently, in a manner analogous with the non-degenerate case, one might introduce the polynomial structures

$$M^s_{\alpha\beta}(nx,my) = \sum_{x_iy_i} \sqrt{g_{n_1+x_1,m_1+y_1}g_{n_2+x_2,m_2+y_2}\cdots g_{n_r+x_r,m_r+y_r}}\,\Pi^s_{\alpha\beta}(nxx_i,myy_i).$$

Quantities $\Pi^s_{\alpha\beta}(nxx_i,myy_i)$ appear to be polynomials only from a computational point of view. The expressions following from this perturbation theory have hence a polynomial form after the calculation of the corresponding matrix elements; through the initial coefficients $U^n_{xx_i}$ and because, within the group of degenerate states, we must retain $E^\sigma_{nx} - E^\sigma_{ny}$ rather than $E^0_{nx} - E^0_{ny}$ in the denominators of expansions, the dependences of $\Pi^s_{\alpha\beta}(nxx_i,myy_i)$ on the quantum numbers might be more complicated than merely of polynomial form. For this reason, it is preferable to derive the principal recurrence relations not for polynomials but rather for matrix elements $M^s_{\alpha\beta}(nx,my)$, from which the polynomial structures become readily reconstructed according to the above definition.

With regard to the equations for matrix elements, it suffices to multiply expression (3.8) for $|nx,\alpha\rangle$ by $\langle \ell z,\mu|\xi_1^{s_1}\xi_2^{s_2}\cdots\xi_r^{s_r}$ on the left. As a result, we obtain

$$\begin{aligned} M^s_{\mu\alpha}(\ell z,nx) &= \frac{1}{\alpha}\sum_{y;m\neq n}\sum_{(pq\beta\gamma\nu)\alpha} p \sum_{(j)p+2} a_j\Delta^0_q(nx,my)M^s_{\mu\nu}(\ell z,my)M^j_{\beta\gamma}(my,nx) \\ &\quad + \frac{1}{\alpha}\sum_{y\neq x}\sum_{(pq\beta\gamma\nu)\alpha>\sigma} p \sum_{(j)p+2} a_j\Delta^\sigma_q(nx,ny)M^s_{\mu\nu}(\ell z,ny)M^j_{\beta\gamma}(ny,nx), \end{aligned} \tag{3.9}$$

in which indices j and s denote, as before, integers in sets $\{j_1,j_2,\ldots,j_r\}$ and $\{s_1, s_2,\ldots,s_r\}$, respectively. These general recurrence relations determine an arbitrary matrix element in the presence of degeneracy. The non-degenerate case is clearly contained here for $\sigma = 0$. One can thus reconstruct all elements $M^s_{\alpha\beta}(nx,my)$ and, assuming $\lambda = 1$ in final formulae, calculate the eigenvalues of an anharmonic Hamiltonian H,

$$E_{nx} = E^0_n + \sum_\alpha \frac{1}{\alpha}\sum_{(p\beta\gamma)\alpha} p \sum_{(j)p+2} a_j M^j_{\beta\gamma}(nx,nx)$$

and the matrix elements of a function f,

$$(nx|f|my) = \sum_{\ell\beta\gamma}\sum_{(s)\ell} \frac{2^{-\ell/2}}{\ell!} f^{(\ell)}_s M^s_{\beta\gamma}(nx,my),$$

of which the explicit form is determined according to the expression,

$$(nx|f|my) = \sum_{x_i y_i} \sqrt{g_{n_1+x_1,m_1+y_1} g_{n_2+x_2,m_2+y_2} \cdots g_{n_r+x_r,m_r+y_r}} \Phi(nxx_i, myy_i),$$

$\Phi(nxx_i,myy_i)$ are in turn functions of quantum numbers $n_1, n_2,\ldots, n_r$ and $m_1, m_2,\ldots, m_r$. In addition, Φ depends trivially on numbers x_i and y_i, which are zero in the absence of degeneracy ($x_i = y_i = 0$).

The problem of degenerate states is nearly solved; it remains only to mention the selection rule. Quantities $\Pi^s_{\alpha\beta}(nxx_i, myy_i)$ are non-zero provided that the parities of $\sum_i[(n_i + x_i) - (m_i + y_i) + s_i]$ and $\alpha + \beta$ coincide. This condition follows from the selection rule for the polynomials in the non-degenerate case [16] upon replacements $n_i \to n_i + x_i$ and $m_i \to m_i + y_i$.

Advantages of the New Technique

Summarizing the results obtained, we consider the principles of the new formalism beginning from the final postulates [18,20]. As functions of quantum numbers n_1, $n_2,\ldots, n_r$, the polynomials $\Pi^s_{\alpha\beta}(n, n \pm k)$, with indices k and s denoting integers in sets $\{k_1,k_2,\ldots,k_r\}$ and $\{s_1,s_2,\ldots,s_r\}$, are expressible from the system of recurrence relations,

$$\Pi^s_{\alpha\mu}(n, n \pm k) = \frac{1}{\alpha} \sum_{(pq\beta\gamma\nu)\alpha} p \sum_{(j)p+2} a_j \underset{m}{\mathbf{S}'} \{\pm k\} \Delta_q(n,m) \Pi^j_{\beta\gamma}(m,n) \Pi^s_{\nu\mu}(m, n \pm k),$$

the addition theorem, which is especially useful for $\alpha + \beta = \text{const.}$,

$$\Pi^{s+q}_{\alpha\beta}(n, n \pm k) = \underset{m}{\mathbf{S}}\{\pm k\} \Pi^s_{\alpha 0}(n,m) \Pi^q_{0\beta}(m, n \pm k),$$

and the identity relating the polynomials with zero superscripts,

$$\sum_{(\beta\gamma)\alpha > 0} \Pi_{\beta\gamma}(n, n \pm k) = 0.$$

Here, $\Delta_0(n,m) = (E^0_n - E^0_m)^{-1}$, and

$$\Delta_q(n,m) = \frac{1}{E^0_n - E^0_m} \sum_i \sum_{(\alpha_1\alpha_2\cdots\alpha_i)q} \prod_{e=1}^{i} \frac{E^{\alpha_e}_m - E^{\alpha_e}_n}{E^0_n - E^0_m}.$$

Inspection of the first summation in the recurrence relations shows that, in accordance with the equality $p + q + \beta + \gamma + \nu = \alpha$, the indices q, β, γ and ν take values from 0 to $\alpha - 1$, whereas $p = 1, 2, \ldots, \alpha$. The subscripts of the polynomials on the right side thus apparently do not exceed $\alpha - 1$. As the subscripts determine the order of the polynomial in λ, we obtain the solution for arbitrary polynomials $\Pi^s_{\alpha\mu}(n, n \pm k)$ beginning with $\alpha = 1$ and $\mu = 0$. In addition, unlike the detailed form of the recurrence relations in Ref. [16], we introduce here a special formalistic symbol of summation

$$\underset{m}{\mathbf{S}}\{\pm k\} = \prod_{i=1}^{r} \underset{m_i}{\mathbf{S}}\{\pm k_i\},$$

in which

$$\underset{m_i}{\mathbf{S}}\{+k_i\} = \sum_{m_i < n_i} g_{m_i n_i} + \sum_{n_i \le m_i \le n_i + k_i} + \sum_{m_i > n_i + k_i} g_{n_i + k_i, m_i},$$

and, for $m_i \ge n_i$, the factor $g_{n_i m_i}$ is equal to $m_i!/n_i!$; otherwise $g_{n_i m_i} = n_i!/m_i!$. If an expression contains quantity $g_{n_i m_i}$ with $m_i < n_i$, $g_{n_i m_i}$ should hence be replaced with $g_{m_i n_i}$. Symbol

$$\underset{m_i}{\mathbf{S}}\{-k_i\} = \sum_{m_i > n_i} g_{n_i m_i} + \sum_{n_i - k_i \le m_i \le n_i} + \sum_{m_i < n_i - k_i} g_{m_i, n_i - k_i}$$

has the same interpretation. The prime on the summation symbol indicates that the term with $m = n$ $(m_i = n_i)$ should be omitted. The quantities E_n^α appearing in the factor $\Delta_q(n,m)$ are corrections of order λ^α to harmonic-oscillator energy E_n^0, so that the eigenvalues of H become

$$E_n = E_n^0 + \sum_\alpha E_n^\alpha.$$

The denominator of the expression for $\Delta_q(n,m)$ contains only differences $E_n^0 - E_m^0$ to varied degree, which evidently produce no dependence on n after summation over m. This condition proves that the quantities in question are polynomials of quantum numbers $n_1, n_2, \ldots, n_r$. The polynomials are symmetric under simultaneous permutations of the subscripts and all pairs of quantum numbers:

$$\Pi^s_{\alpha\beta}(n, n \pm k) = \Pi^s_{\beta\alpha}(n \pm k, n).$$

The polynomials with zero subscripts $\Pi^s(n, n \pm k)$ are readily calculated with the aid of the addition theorem or simply through the matrix element [19,20].

Having reconstructed the desired polynomials, one might define the eigenvalues of the Hamiltonian H,

$$E_n = E_n^0 + \sum_\alpha \frac{1}{\alpha} \sum_{(p\nu)\alpha} p \sum_{(j)p+2} a_j \Pi^j_{(\beta\gamma)\nu}(0),$$

and the matrix elements of a function f,

$$(n|f|n+k) = \sqrt{g_{n,n+k}} \sum_{\ell\alpha} \sum_{(s)\ell} \frac{2^{-\ell/2}}{\ell!} f_s^{(\ell)} \Pi^s_{(\beta\gamma)\alpha}(k),$$

in which $g_{n,n+k} = g_{n_1,n_1+k_1} g_{n_2,n_2+k_2} \cdots g_{n_r,n_r+k_r}$. It is convenient to tabulate not the polynomials themselves but their convolutions, i.e. the sums of polynomials of the same order:

$$\Pi^s_{(\beta\gamma)\alpha}(\pm k) \equiv \sum_{(\beta\gamma)\alpha} \Pi^s_{\beta\gamma}(n, n \pm k).$$

The convolution operation decreases the highest degree in n and is applied to polynomials $\Pi^s_{\beta\gamma}(n_1, n_1+k_1; n_2-k_2, n_2; n_3, n_3-k_3, \ldots)$ with arbitrary values of k. Note here that invariably

$$k_i \le k_{\max} = 3(\beta+\gamma) + \sum_i s_i,$$

and the polynomials vanish for $k_i > k_{\max}$ [16].

In the general case, to calculate the observable quantities, one must use Eq. (3.9), which allows one to take into account the degenerate levels. From a practical point of view, for both one-dimensional and many-dimensional problems, Eq. (3.9) is principal; having obtained with its aid all matrix elements and the result divided by factor $\sqrt{g}$, we reconstruct polynomial structures. One should express matrix elements in a polynomial manner and tabulate not the polynomials but their convolutions. Although we can work with equations of polynomials, relations (3.9) are convenient and simple for the calculation of higher-order approximations in perturbation theory. The equations with polynomials are necessary to exhibit and to prove the polynomial structure of Π-quantities in explicit form.

We consider briefly an example of the construction of a formalism for the one-dimensional case. The initial data are harmonic polynomials $\Pi^s(n,m)$, which are found according to the addition theorem or trivial calculations of matrix elements $\langle n|\xi^s|m\rangle$:

$$\Pi^s(n, n+s) = 1, \Pi^2(n,n) = 2n+1, \text{ etc.}$$

The influence of anharmonicity is described with the system of functions $G_p(\xi)$; as $r = 1$, it is convenient to choose G_p in form $\hbar\omega a_p \xi^{p+2}$, in which ω and ξ are defined above, and coefficients a_p are dimensionless. The first polynomials in this case are linear in a_1 and follow from the recurrence equations. For instance,

$$\Pi^3_{01}(n,n) = \Pi^3_{10}(n,n) = -a_1(30n^2 + 30n + 11).$$

The convolution $\Pi^3_{(\beta\gamma)1}(0)$ together with $\Pi^4(n,n)$ clearly forms the energy in the second-order approximation

$$E_n \approx E_n^0 + E_n^1 + E_n^2, \text{ in which } E_n^1 \equiv 0.$$

Second-order polynomials are determined in an analogous manner. We initially calculate Π_{02}; from identity $\Pi_{(\beta\gamma)2}(\pm\, k) = 0$, we then find Π_{20}, and, eventually, with the addition theorem, we reconstruct Π^s_{02} and Π^s_{20}. This procedure is repeated for approximations of third and greater orders.

In conclusion of this section, we discuss a numerical value of the highest degree in quantum number of the polynomial. According to Eq. (3.5),

$$\Pi^s_{\alpha\beta}(n, n+k) = \left(\sqrt{g_{n,n+k}}\right)^{-1} \langle n, \alpha | \xi^s | n+k, \beta \rangle.$$

After quantization, variable ξ practically converts into $\sqrt{n}$, so that $\xi^s \rightarrow n^{s/2}$. Vector $|n,\alpha\rangle$ is formed primarily by quantity $\xi^{3\alpha}$, and $|n+k,\beta\rangle$ analogously by $\xi^{3\beta}$; therefore,

$$\Pi^s_{\alpha\beta}(n, n+k) \sim n^{(s+3\alpha+3\beta-k)/2}.$$

Taking into account that $3(\alpha+\beta) + s = k_{\max}$, we find for the highest degree in n of both polynomials $\Pi^s_{\alpha\beta}(n, n+k)$ and $\Pi^s_{\beta\alpha}(n-k, n)$ the value $(k_{\max} - k)/2$ with $k \neq 0$. If $k = 0$ and $\alpha + \beta$ is an odd number, the resultant degree decreases by unity and becomes equal to $(k_{\max}/2) - 1$. We arrive at this conclusion readily if we take into account, for example, that $k_{\max} \geq 2$ in this case. For an even value of $\alpha + \beta$, when $k = 0$, the highest degree in n is simply equal to $k_{\max}/2$.

As an illustration, we write two polynomials

$$\Pi^3_{02}(n, n+5) = \frac{a_1^2}{12}(80n^2 + 495n + 639) + \frac{a_2}{4}(11n + 39) \text{ and}$$
$$\Pi^3_{20}(n, n+5) = \frac{a_1^2}{12}(80n^2 + 465n + 549) - \frac{a_2}{4}(11n + 27),$$

add to them this one

$$\Pi^3_{11}(n, n+5) = -\frac{a_1^2}{3}(40n^2 + 240n + 234),$$

and calculate the convolution $\Pi^3_{(\beta\gamma)2}(5)$. Clearly, $\Pi^3_{(\beta\gamma)2}(5) = 21a_1^2 + 3a_2$; the highest degree $(k_{\max} - k)/2$ equal here to two decreases, and the sought result becomes much simpler.

Polynomials and Computational Rules

First, the polynomials form, with the required accuracy, all necessary physical observables of the anharmonicity problem. The desired quantities are obtained immediately on solving or opening the recurrence equations or relations avoiding conventional intermediate manipulations. We compare two schemes to construct the stationary perturbation theory:

1. Schrödinger equation→eigenfunctions and eigenvalues→matrix elements;
2. recurrence equations→eigenvalues and matrix elements.

The first scheme is conventional, whereas we proposed the second scheme. The main disadvantage of the conventional scheme is that, at each stage, one must return virtually to the beginning – to the Schrödinger equation – to improve the eigenfunctions by increasing the order of the perturbation calculation. Only after these calculations, one is in a position to evaluate the matrix elements. In our method, intermediate calculations are performed on an equal footing, i.e. the procedures to calculate the eigenvalues and arbitrary matrix elements are performed simultaneously.

Second, the proposed theory automatically keeps track of non-zero contributions of the total perturbation to the result sought (see the selection rule below), and takes into account the history of the calculations, i.e. the intermediate calculations. This advantage is achieved on expanding, in a small parameter, the derivatives of the energies and their wave functions, rather than by expanding the eigenfunctions and eigenvalues as is done traditionally. In this sense, the expansion in exact eigenvectors plays a principal role [16],

$$\frac{\partial}{\partial\lambda}|n,\lambda) = \sum_{m\neq n} C_{mn}|m,\lambda),$$

because it ensures a full use of the history of the calculations and, consequently, significantly simplifies the general solution algorithm. If the expansion is performed in terms of the exact eigenvectors, rather than in terms of zero-order basis functions, it is assumed that the former functions exist and are expressible algebraically, for example, with recurrence relations. In addition, one might avoid the renormalization of the function; this problem presents considerable difficulties in the traditional approach in which the function should be renormalized upon passing from one perturbation order to the next.

Other advantages of this method appear in various applications of this perturbation theory. For example, in a framework of the polynomial formalism, one might consider the problem of electro-optical anharmonicity; this problem involves an electric dipolar-moment function d in a non-linear form, and its solution requires evaluation of matrix elements $(n|d|m)$. The absolute values of dipolar-moment derivatives $d^{(s)}$ might be unknown beforehand, which complicates the problem. In the traditional formalism, the consideration proceeds, as a rule, from the wave function of a definite order, which leads to the loss of significant contributions. In the polynomial formalism, we consider separately each term in an expansion of the dipolar-moment function and, consequently, calculate the entire matrix element in a given order in a small parameter. For instance, let

$$d(q) = d^0 + d'q + \frac{d''}{2}q^2 + \frac{d'''}{6}q^3.$$

If the anharmonicity is such that d'' is ~10 times d' and ~100 times d''', a conventional calculation of matrix element $(n|d|m)$ in the second order of perturbation theory yields

$$(n|d|m) = d^0\delta_{nm} + \underbrace{d'(n|q|m) + \frac{d''}{2}(n|q^2|m) + \frac{d'''}{6}(n|q^3|m)}_{\text{second order}}.$$

This result is incorrect, however, because the matrix element is a sum of terms of disparate orders. Namely, $d'(n|q|m)$ is a third-order quantity and $d'''(n|q^3|m)$ is a fourth-order quantity. To improve this situation, one must calculate in a somewhat different manner (see below the rule of order):

$$(n|d|m) = d^0\delta_{nm} + \underbrace{d'(n|q|m)}_{\text{first order}} + \underbrace{\frac{d''}{2}(n|q^2|m)}_{\text{second order}} + \underbrace{\frac{d'''}{6}(n|q^3|m)}_{\text{zero order}}.$$

This approach is especially simple to implement in a polynomial formalism. What is necessary in the above case is to evaluate two convolutions $\Pi^2_{(\alpha\beta)2}$ and $\Pi^1_{(\alpha\beta)1}$, and to reconstruct the harmonic polynomial Π^3.

Summarizing the above analysis, the observable intensities and frequencies of molecular transitions are associated physically with matrix elements. Frequencies are associated with differences of diagonal matrix elements of a Hamiltonian, but intensities with matrix elements of dipolar-moment function d,

$$(n|d|n+k) = \sqrt{g_{n,n+k}} \sum_{\ell\alpha} \sum_{(s_1 s_2 \cdots s_r)\ell} \frac{2^{-\ell/2}}{\ell!} d^{(\ell)}_{s_1 s_2 \cdots s_r} \Pi^{s_1 s_2 \cdots s_r}_{(\beta\gamma)\alpha}(k).$$

Because the quantum-mechanical amplitude λ is typically $\sim 10^{-1}$, the expansion coefficients of the dipolar-moment function $d^{(\ell)}_s$ can be assumed to be proportional to λ^{σ_s}, in which $\sigma_{s_1 s_2 \cdots s_r}$ is an integer that determines the order of $d^{(\ell)}_{s_1 s_2 \cdots s_r}$ in terms of λ. If electro-optical effects are weak, $\sigma_s = \ell$; hence, the difference between σ_s and ℓ characterizes the strength of electro-optical anharmonicity. This condition becomes a definition of electro-optics. Expanding the dipolar-moment function in terms of vibrational variables,

$$d = \sum_{\ell} \sum_{(s)\ell} \frac{2^{-\ell/2}}{\ell!} d^{(\ell)}_{s_1 s_2 \cdots s_r} \xi_1^{s_1} \xi_2^{s_2} \cdots \xi_r^{s_r},$$

we obtain automatically the dependence of the matrix element on the quantum-mechanical amplitude. Quantity $\xi_i^{s_i}$ is associated with λ^{s_i}; although it might seem that $d_s^{(\ell)} \sim \lambda^{\ell}$, this association is incorrect. The behaviour of derivatives $d_s^{(\ell)}$ can deviate strongly from that of λ^{ℓ}, which indicates the presence of another (electro-optical) nature of anharmonicity, as distinct from the mechanical anharmonicity related to the Hamiltonian. The greater the difference between σ_s and ℓ, the stronger the electro-optical anharmonicity; the equality $\sigma_s = \ell$ is indicative of the absence of the latter. One might apply this analysis to an arbitrary coordinate function *f*, of physical interest, in an analogous manner.The general rules pertinent for calculating the matrix elements are [18]:

1. for the matrix element to be represented in the same order of a small parameter, it suffices to satisfy the equality $\sigma_{s_1 s_2 \cdots s_r} + \alpha = \text{const.}$ (the rule of order);
2. for convolutions $\Pi^{s_1 s_2 \cdots s_r}_{(\beta\gamma)\alpha}(k)$ to be non-zero, numbers $\sum_i (k_i - s_i)$ and α have the same parity; otherwise, the polynomials vanish identically (the selection rule).

With respect to the rule of order, some comments appear above; here we consider the selection rule. Let the value of sum $\sum_i k_i$ be odd; the contributions with even ℓ should then be taken into account for odd values of α, and the contributions with odd values of ℓ are associated with even values of α. Conversely, if the value of $\sum_i k_i$ is even, both ℓ and α are either even or odd. Through the selection rule, half of all possible contributions of the perturbation to an arbitrary matrix element vanish. The same condition is true for eigenvalues E_n. As $E_n \sim \Pi^{j_1 j_2 \cdots j_r}_{(\beta\gamma)\nu}(0)$, the sum of all j_i and ν should be even; the other variants result in zero. This polynomial technique possesses a pronounced structure: all necessary quantities are directly determined in terms of non-zero polynomials or through their convolutions, which can be tabulated to facilitate calculations.

Electro-Optics of Molecules

In the preceding section, we consider a correct calculation of the matrix elements of a model dipolar-moment function,

$$d(q) = d^0 + d'q + \frac{d''}{2}q^2 + \frac{d'''}{6}q^3,$$

in which d'' is ~ 10 times d' and ~ 100 times d'''. Despite the abstraction, this simple model is applicable to describe an anomalous intensity distribution for real molecules, e.g. diatomic defects OD in ionic crystals NaBr, KF and NaI. According to Ref. [41], for OD in these crystal matrices, the intensity of the second harmonic is ~ 10 times that of the first harmonic, which is inexplicable in terms of only mechanical anharmonicity [19,42].

The effects of electro-optical anharmonicity are various. For instance, the intensities of vibration–rotational transitions in absorption are associated with derivatives of *d*; the first derivative generally plays a prominent role, but in atypical cases other derivatives also appear to strongly influence these intensities. A variation of the value d' can exert an influence in electro-optical phenomena. In systems consisting, typically, of molecules with a large first derivative of dipolar moment with respect to a normal coordinate, the resonance dipole–dipole interaction begins to play a dominant role. The mechanism of this interaction is most conspicuous in the IR absorption spectra of ionic crystals that contain impurity defects XH (X = O, S, Se, Te) [43,44], and in spectra of low-temperature liquids SF_6, CF_4, NF_3 and OCS [45]. In both cases, through intramolecular interaction, shifts in vibrational levels and substantially altered intensities of transitions are observed.

Phenomenon of Strong Anharmonicity

To illustrate how the order and the selection rules work, we consider a calculation of matrix elements for a diatomic molecule with dipolar-moment function d,

$$d(\xi) = \sum_s D_s \xi^s, \quad D_s = \frac{2^{-s/2}}{s!} d^{(s)}.$$

According to Eq. (3.7),

$$(n|d|n+k) = \sqrt{g_{n,n+k}} \sum_{s\alpha} D_s \Pi^s_{(\beta\gamma)\alpha}(k).$$

Indices s and α, in which $\alpha = \beta + \gamma$, clearly denote the orders of the electro-optical and mechanical approximations, respectively. We investigate the exact second order [16]. In this case,

$$(n|d|n+k) = \sqrt{g_{n,n+k}} \sum_{s=0}^{3} D_s \sum_{\alpha=0}^{2} \Pi^s_{(\beta\gamma)\alpha}(k).$$

Having simply calculated the first polynomials (see section 'Background'), we obtain the expectation value of the dipolar moment of a diatomic molecule in state $|n)$:

$$(n|d|n) = d^0 + \frac{1}{8}(7a^2 - 3A)d'' - \frac{7}{48}ad''' + \left(\frac{d''}{2} - 3ad'\right)\left(n + \frac{1}{2}\right) + \Lambda_0\left(n + \frac{1}{2}\right)^2,$$

and the matrix element responsible for the fundamental transition,

$$(n|d|n+1) = \sqrt{\frac{g_{n,n+1}}{2}}\left(d' + \frac{1}{32}(31a^2 - 14A)d''' + \left(\frac{1}{4}(11a^2 - 6A)d' - \frac{5}{2}ad'' + \frac{1}{4}d'''\right)(n+1) + \Lambda_1(n+1)^2\right),$$

in which

$$\Lambda_0 = \frac{3}{2}(5a^2 - A)d'' - \frac{5}{4}ad''' \text{ and } \Lambda_1 = \frac{1}{32}(173a^2 - 34A)d'''.$$

Here, instead of a_p, we introduce traditional parameters that are equal to

$$2^{(p+2)/2}a_p,$$

in particular, $a = 2\sqrt{2}a_1$ and $A = 4a_2$. Moreover, we imply that $\lambda = 1$; thus, $\lambda^p a_p \to a_p$ and the real order in quantum-mechanical amplitude λ is included in a_p.

In the matrix elements derived, we take into account electro-optical coefficients Λ_0 and Λ_1, and we retain corrections

$$(7a^2 - 3A)d''/8, \ (-7/48)ad''' \text{ and } (31a^2 - 14A)d'''/32$$

to d^0 and d', respectively. These coefficients and corrections correspond to a condition in which the dipolar-moment function deviates considerably from a linear behaviour; they express an influence of a strong electro-optical anharmonicity [16]. These coefficients and corrections must be ignored if the first derivative of the dipolar moment has a large value. In the next order of polynomial perturbation theory, polynomials $\Pi^1_{03}(n,n)$ and $\Pi^1_{30}(n,n)$ contribute to electro-optical coefficient Λ_0; the latter is especially important at large values of d'. Through the order rule, only polynomial $\Pi^4(n,n)$ affects coefficient Λ_0, and polynomials $\Pi^4_{01}(n,n+1)$ and $\Pi^4_{10}(n,n+1)$ make a contribution to Λ_1 at small values of first derivative d'.

A similar treatment is valid for higher harmonics. For pertinent matrix elements, we have

$$(n|d|n+2) = \frac{1}{4}\sqrt{g_{n,n+2}}\left(2ad' + d'' + \Lambda_2\left(n + \frac{3}{2}\right)\right) \text{ and}$$

$$(n|d|n+3) = \frac{1}{4}\sqrt{\frac{g_{n,n+3}}{2}}\left(\left(\frac{3}{2}a^2 + A\right)d' + 2ad'' + \frac{1}{3}d''' + \Lambda_3(n+2)\right),$$

in which $\Lambda_2 = \frac{1}{4}(a^2 - 10A)d'' - 2ad'''$ and $\Lambda_3 = -(2a^2 + A)d'''$.

We apply these formulae to hydrogen iodide (HI); through its unique electrical properties, this molecule attracts special attention [46–48]. According to Ref. [48], the values of the first three derivatives of the dipolar-moment function for HI have comparable magnitudes:

$$d' = -0.00552, \quad d'' = 0.00568, \quad d''' = -0.00505.$$

Here and after, we express the dipolar moment and its derivatives in debye units. We emphasize again that $\lambda = 1$; thus $\lambda^s d^{(s)} \to d^{(s)}$ and the real order in λ is included in $d^{(s)}$. The force parameters for HI have standard values [49]:

$$a = -0.095 \text{ and } A = 0.0103.$$

On substituting these parameters into the expressions for the matrix elements and with elementary manipulations, we obtain

$$(0|d|0) = 0.44551(0.44541) \text{ and } (0|d|1) = -0.004029(-0.004016);$$

also

$$(0|d|2) = 0.001800(0.001804) \text{ and } (0|d|3) = -0.001129(-0.001124),$$

in which the numbers within parentheses are the experimental values (see references in [12]). With the aid of this example of HI, we readily understand the physical meaning of electro-optical coefficients Λ. For this purpose, we disregard the aforementioned corrections to quantities d^0 and d' in relations for $(n|d|n)$ and $(n|d|n+1)$, and set $\Lambda_i = 0$, for $i = 0$, 1, 2 and 3. As a result, we obtain these matrix elements,

$$(0|d|0) = 0.4456,\quad (0|d|1) = -0.00388,\quad (0|d|2) = 0.00237,\quad (0|d|3) = -0.00127.$$

Compared together, they show that the discrepancy between values calculated here and those of the matrix element $(0|d|0)$ above is insignificant; one explanation is that the expectation value of the dipolar moment almost coincides with the static dipolar moment, $d^0 = 0.445$ [12]. The error in determining matrix element $(0|d|1)$ is small, but the relative error in calculating higher harmonics attains 30% (see transition $0 \rightarrow 2$). This analysis confirms the necessity of including the electro-optical coefficients in the calculation.

An inherently strong electro-optical anharmonicity involves additional exclusive harmonics with matrix elements

$$(n|d|n+4) = \frac{\sqrt{g_{n,n+4}}}{8}\left(\left(\frac{5}{2}a^2 + A\right)d'' + ad'''\right) \text{ and}$$

$$(n|d|n+5) = \sqrt{\frac{g_{n,n+5}}{2}}\left(\frac{7}{2}a^2 + A\right)\frac{d'''}{16}.$$

For HI, we obtain

$$(0|d|4) = 0.000408(0.000395) \text{ and } (0|d|5) = -0.0001024(-0.000136);$$

the numbers within parentheses are experimental values of these matrix elements taken from Ref. [48]. The matrix elements of the exclusive harmonics for the HI molecule are a tenth of those for the matrix elements of the first overtones. As is easily seen, the exclusiveness resides not in the existence of such high harmonics by themselves but in their appearance with the first harmonics already present in the second order of perturbation theory.

The Direct and Inverse Problems of Spectroscopy

An improved application of perturbation theory for the problem of the electro-optical anharmonicity of molecules has renewed interest in a well-known problem. Among various molecules, HI is of special interest because, for this molecule, together with the first harmonics, the fourth and fifth harmonics play important roles. To illustrate the power of the polynomial formalism, we consider the solution of the direct problem for HI taking into account the higher-order approximations of perturbation theory [50]. It is convenient to use quantum-chemical values of the derivatives of the dipolar-moment function as initial data [51]:

$$d^0 = 0.4412, \quad d' = -7.16 \cdot 10^{-3}, \quad d'' = 6.62 \cdot 10^{-3}, \quad d''' = -5.39 \cdot 10^{-3},$$

$$d^{\mathrm{IV}} = -8.92 \cdot 10^{-5}, \quad d^{\mathrm{V}} = 2.84 \cdot 10^{-4}, \quad d^{\mathrm{VI}} = 3.03 \cdot 10^{-4}.$$

These quantities, which are obtained in a framework of the DC-CCSD(T) method, represent the first seven coefficients of the expansion for the dipolar moment with respect to the normal coordinate. The fifth order being sufficient for the present purpose, we write the necessary force parameters [52]:

$$a = \lambda a_1^D/2 = -9.56 \cdot 10^{-2}, \quad A = \lambda^2 a_2^D/2 = 1.14 \cdot 10^{-2}, \quad b = \lambda^3 a_3^D/2 = -1.16 \cdot 10^{-3},$$

$$B = \lambda^4 a_4^D/2 = 1.12 \cdot 10^{-4}, \quad c = \lambda^5 a_5^D/2 = -1.14 \cdot 10^{-5}, \quad \lambda = \sqrt{2B_e/\nu_e} = 0.0751,$$

in which B_e denotes the equilibrium rotational parameter (in cm^{-1}) and ν_e is the wavenumber (in cm^{-1}) of harmonic vibration. We note a simple relation between polynomial coefficients a_p and Dunham's parameters a_p^D,

$$a_p = 2^{-(p+4)/2} a_p^D \lambda^p.$$

The matrix elements are generally defined with this expression,

$$(0|d|k) = \sqrt{k!} \sum_{s=0}^{6} D_s \sum_{\alpha=0}^{5} \Pi^s_{(\beta\gamma)\alpha}(k).$$

Having calculated all non-vanishing polynomial convolutions, we obtain the sought quantities, $(0|d|k)$ (Table 3.1, Π-formalism). Our results [50] agree satisfactorily with previously calculated values [51] and with contemporary experimental data [53].

The polynomial formalism is applicable to an inverse problem, not just to a direct problem of spectroscopy in which observable matrix elements are determined through the anharmonicity parameters. The aim of the inverse problem is to reconstruct the force parameters of potential energy and dipolar-moment coefficients

Table 3.1 Vibrational Matrix Elements/Debye of the Dipolar Moment for HI

Matrix Element	Π-Formalism [50]	Theoretical Results [51]	Experimental Data [53]
$(0\|d\|0)\times 10^1$	4.417	4.416	4.477
$(0\|d\|1)\times 10^3$	− 5.00	− 5.00	− 4.07
$(0\|d\|2)\times 10^3$	2.15	2.15	1.97
$(0\|d\|3)\times 10^3$	− 1.21	− 1.21	− 1.15
$(0\|d\|4)\times 10^4$	4.30	4.29	4.00
$(0\|d\|5)\times 10^4$	− 1.44	− 1.43	− 1.36
$(0\|d\|6)\times 10^5$	5.01	5.07	
$(0\|d\|7)\times 10^5$	− 1.88	− 1.87	

with the aid of experimental data; we must consider the accuracy of the expected results and their interpretation. We reconstruct the force parameters from a_1 to a_4 using formulae for the wavenumbers of vibrational transitions obtained in the fourth order of perturbation theory; we then apply the same procedure but in the sixth order. Do we improve the values of the first four coefficients a_1, a_2, a_3 and a_4? The answer is negative: rather, we attribute the new values to all quantities a_p. As an example, we refer to the case of the HI molecule above; the values for a_1 and a_2 [49] used in the theory at the second order differ from the same quantities at the sixth order [52].

As a result of the inverse problem, the force parameters have physical meaning in only the order in which these parameters are obtained. To evaluate coefficients $a_7, a_8, \ldots$ of greater order, one must, moreover, take into account non-adiabatic corrections to correctly calculate the sought result [6]. If we neglect the non-adiabatic influence and calculate the first 10 parameters from a_1 to a_{10},

$$2^{(p+2)/2} a_p \times 10^p = \text{A.abcdef}\ldots,$$

we can be certain that some significant digits a, b, c,... are simply altered. The quantity a_1 is least affected, but for a_6 all significant figures a, b,.... are suspect. The highest coefficients, beginning from a_7, have no physical meaning at all in the adiabatic approximation [54].

The next step of the inverse problem is a direct calculation of matrix elements of coordinate ξ to various powers,

$$(0|\xi^s|k) = \sqrt{k!}\sum_{\alpha} \Pi^s_{(\beta\gamma)\alpha}(k).$$

This step is readily accomplished if all calculations are made in the order in which the force parameters are reconstructed. In addition, the results are obtained with adequate accuracy because, through the selection rule, half of all contributions of the perturbation to an arbitrary matrix element vanish. For instance, if both k and s are even or odd, the non-vanishing convolutions become

$$\Pi^{s}_{(\beta\gamma)2}(k),\quad \Pi^{s}_{(\beta\gamma)4}(k),\quad \Pi^{s}_{(\beta\gamma)6}(k),\quad \Pi^{s}_{(\beta\gamma)8}(k),\ldots.$$

The result of calculating matrix elements, for example, in the sixth order, is therefore correct with accuracy up to the eighth order. The case in which k and s have distinct parity is analysed in the same manner; the non-vanishing polynomial convolutions are obviously

$$\Pi^{s}_{(\beta\gamma)1}(k),\quad \Pi^{s}_{(\beta\gamma)3}(k),\quad \Pi^{s}_{(\beta\gamma)5}(k),\quad \Pi^{s}_{(\beta\gamma)7}(k),\ldots.$$

The final stage is characterized with a consideration of quantity

$$(0|d|k) = \sum_{s} D_s(0|\xi^s|k);$$

one must generate coefficients D_s through the experimental matrix elements of dipolar moment, commonly with the aid of a convenient numerical method, for instance, least squares. The principal difficulty in the solution of the inverse problem perhaps lies in the uncertainty associated with a direct calculation of various matrix elements. Namely, we have

$$(0|d|k) = \sqrt{k!}\sum_{s} D_s \sum_{\alpha=0}^{?} \Pi^{s}_{(\beta\gamma)\alpha}(k),$$

in which the question mark appears to underline the incorrectness of combining the parameters of the direct and inverse problems in one formula. The point is that the order pertinent to each quantity D_s is unknown; the order of mechanical anharmonicity necessary to calculate matrix element $(0|\xi^s|k)$ is hence also unknown. This relevant condition has previously not been addressed, but to solve the inverse problem, one must take it into account [16]. To satisfy the order rule, one must preliminarily know the relative orders of all quantities D_s. For instance, if the magnitudes of the dipolar-moment derivatives decrease monotonically according to

$$d^{(s)} \sim \lambda^s,$$

to calculate the matrix elements we apply the scheme of decreasing order,

$$(0|d|k) = \sqrt{k!}D_0\delta_{0k} + \sqrt{k!}D_1 \sum_{\alpha=0}^{l} \Pi^{1}_{(\beta\gamma)\alpha}(k) + \sqrt{k!}D_2 \sum_{\alpha=0}^{l-1} \Pi^{2}_{(\beta\gamma)\alpha}(k) + \cdots,$$

in which l denotes the maximum possible value of the order. In other cases, this scheme is invalid.

Extraneous Quantum Numbers

Before this point, we have considered purely vibrational states and neglected possible effects of molecular rotation. Not only does there exist no exact description of the vibration–rotational interaction, but even the separation of molecular motions of vibrational and rotational types is typically conditional and formal. To generalize the theory developed, we consider only one part of this major problem for which the pure rotational problem is solved and the vibrational Hamiltonian contains extraneous rotational quantum numbers as parameters. In this case, the eigenvalue equation has a form

$$(H + \rho\upsilon(N))|nN) = E_{nN}|nN). \tag{3.10}$$

Here, H is the vibrational Hamiltonian, ρ is a function of variable ξ and $\upsilon(N)$ is a function of a combination with extraneous quantum number N. For instance, for diatomic and linear polyatomic molecules, $\upsilon = (1/2)J(J+1)$, in which J is the rotational quantum number. In the case of free librations, the dependence of υ on N might be more complicated. For simplicity, we restrict ourselves to a case of one variable, but one might extrapolate our theory to a many-dimensional case.

One might obtain the solution of Eq. (3.10) in the conventional form of polynomials of quantum numbers if the effective potential including $\rho\upsilon$ is preliminarily expanded in powers of ξ. In this case, the algorithm for deriving corrections to the energy and the wave function is invariant. We then encounter difficulties in calculating matrix elements $(nN|f|n'N')$ non-diagonal in N because they lack a direct expression in terms of polynomials. To avoid this problem, one might apply a simple theorem [18–20].

We begin this section from the proof of this theorem. Matrix element $(nN|f|n'N')$ is formally represented as an expansion in powers of $\Delta N = N' - N$:

$$\begin{aligned}(nN|f|n'N') &= (nN|f \exp(\Delta N\partial/\partial N)|n'N)\\ &= (nN|f|n'N) + (nN|f\ \partial/\partial N|n'N)\Delta N + (nN|f\ \partial^2/\partial N^2|n'N)\frac{\Delta N^2}{2} + \cdots.\end{aligned}$$

Matrix elements $(nN|f|n'N')$ non-diagonal in N are uniquely reduced to diagonal elements on replacing function f with $f\exp(\Delta N\partial/\partial N)$. It remains only to specify the action of derivative $\partial/\partial N$ on vector $|nN)$. For this purpose, we formally differentiate Eq. (3.10) with respect to N:

$$\rho\frac{\partial\upsilon}{\partial N}|nN) + (H + \rho\upsilon)\frac{\partial}{\partial N}|nN) = \frac{\partial E_{nN}}{\partial N}|nN) + E_{nN}\frac{\partial}{\partial N}|nN). \tag{3.11}$$

Because the extraneous quantum number appears in the initial equation as a parameter,

$$\frac{\partial}{\partial N}|nN) = \sum_{m\neq n} O_{mn}|mN),$$

in which $O_{nn} = (nN|\partial/\partial N|nN) = 0$ in view of

$$\frac{\partial}{\partial N}(nN|nN) = \frac{\partial}{\partial N}1 = 0.$$

On substituting the expansion for the first derivative of vector $|nN)$ into Eq. (3.11) and taking Eq. (3.10) into account, we obtain coefficients O_{mn}, and hence $\partial/\partial N|nN)$:

$$\frac{\partial}{\partial N}|nN) = \frac{\partial v}{\partial N}\sum_{m\neq n}\frac{(nN|\rho|mN)}{E_{nN}-E_{mN}}|mN). \tag{3.12}$$

This elegant result is exact. The matrix elements of function ρ are diagonal in N; the matrix element of a function f in the first order in ΔN is thus expressible in terms of polynomials of quantum numbers. Subsequent differentiations of expression (3.12) lead to $\partial^2/\partial N^2|nN)$, $\partial^3/\partial N^3|nN)$, etc. Although this expansion in powers of ΔN is somewhat formal, the expression for $(nN|f|n'N')$ after substitution of Eq. (3.12) into it adopts the meaning of an ordinary matrix element and is no longer formal, which leads us to the following useful theorem.

Theorem. For a coordinate function f, one might pass from the matrix elements non-diagonal in the extraneous quantum number to the diagonal matrix elements according to

$$(nN|f|n'N') = (nN|f\ \exp(\Delta N\ \partial/\partial N)|n'N),$$

in which the action of operator $\partial/\partial N$ on vector $|nN)$ is given by formula (3.12).

With the aid of this theorem, one might also evaluate the diagonal matrix elements. This approach applies to the exact energy (the diagonal matrix element of the Hamiltonian), which is conveniently determined from expression (3.11):

$$\frac{\partial E_{nN}}{\partial N} = \frac{\partial v}{\partial N}(nN|\rho|nN).$$

For instance, for the case of freely rotating diatomic molecules, we have $v = (1/2)J(J+1) \equiv K$ and

$$\frac{\partial E_{nK}}{\partial K} = (nK|\rho|nK).$$

Factorization of the Matrix Elements

The problem of determining the dipolar-transition intensities for vibration–rotational transitions of diatomic molecules is associated with the calculation of coefficients of the Herman–Wallis factor $C_{nn'}$, $D_{nn'}$, $E_{nn'}$,... [24], which appear upon expansion of the squared matrix element of the dipolar-moment function d in powers of ΔK:

$$|(nK|d|n'K')|^2 = (n|d|n')^2(1 + C_{nn'}\Delta K + D_{nn'}\Delta K^2 + E_{nn'}\Delta K^3 + \cdots),$$

$$\Delta K = (1/2)(J'(J' + 1) - J(J + 1)),$$

in which J and n are the rotational and vibrational quantum numbers, respectively; $(n|d|n')$ is a purely vibrational matrix element. This factor emphasizes an influence of the anharmonicity of the internuclear potential not only on the wave function but also on the matrix elements. The direct calculation, which uses, for instance, the hypervirial result [6,55,56], consists of a calculation of approximate wave functions with subsequent determination of the matrix elements of the dipolar-moment function. This traditional way of solving the problem [6] hence involves seeking coefficients in the form of expansions in $d^{(s)}/(n|d|n')$, in which $d^{(s)}$ denote derivatives of the dipolar moment with respect to the normal coordinate. Might the factor depend on the choice of dipolar-moment derivatives $d^{(s)}$? With a certain degree of accuracy, yes, it must; in the general case, no, it must not. The electro-optics directly affect the values of $d^{(s)}$, but only indirectly affect the Herman–Wallis factor. An alternative approach to solve the problem [18,20] is to seek the coefficients in a form of series in $(n|d|m)/(n|d|n')$, in which $m \neq n'$. As the values of matrix elements $(n|d|n + k)$ decrease monotonically with increasing k for most molecules, this method is clearly preferable to the traditional procedure. We consider this alternative method in detail.

Taking free rotation into account, the wave equation for a diatomic molecule has a conventional form

$$(H + \hbar\omega\rho_r K)|nK) = E_{nK}|nK).$$

Here, $H = H^0 + \hbar\omega\sum_{p>0}\lambda^p a_p \xi^{p+2}$ is the vibrational Hamiltonian and

$$\rho_r = (2B_e/\nu_e)(1+\lambda q)^{-2} = (2B_e/\nu_e)\sum_{i=0}(1 + i)(-\lambda q)^i,$$

in which B_e is the rotational parameter and ν_e is the wavenumber of harmonic vibration. As indicated above, the solution of this equation for vector $|nK)$ and energy E_{nK} can be represented in terms of polynomials of quantum numbers with coefficients a_p, frequency ω and equilibrium internuclear separation being functions of K. This condition is, however, inapplicable for transition-dipolar moments $(nK|d|n'K')$ when the rotational quantum number alters. The structures of quantities $|nK)$ and $|n'K')$ remain the same; the principal distinction between them is in the force parameters appearing in these functions: in one case, the argument J is primed but is unprimed in the other case. One might understand the correctness of the matrix-element expansion in the Herman–Wallis form; one takes $K' = K + \Delta K$ into account and represents approximately $a_p(K')$ in a form

$$a_p(K) + a_p'(K)\Delta K + a_p''(K)\Delta K^2 + \cdots,$$

in which $a_p'(K)$, $a_p''(K)$, ... are coefficients. As a result,

$$(nK|d|n'K') = (nK|d|n'K) + (\ldots)\Delta K + (\ldots)\Delta K^2 + \cdots.$$

To show this approach exactly, we formally expand the matrix element of the dipolar-moment function in powers of ΔK:

$$(nK|d|n'K') = (nK|d\ \exp(\Delta K\ \partial/\partial K)|n'K)$$
$$= (nK|d|n'K) + (nK|d\ \partial/\partial K|n'K)\Delta K + (nK|d\ \partial^2/\partial K^2|n'K)\frac{\Delta K^2}{2} + \cdots,$$

and make use of the theorem of extraneous quantum numbers (3.12) in which $v = K$. The sought expansion has a form

$$(nK|d|n'K') = (n|d|n') + \hbar\omega \sum_{m\neq n'} (n|d|m)\frac{(m|\rho_r|n')}{E_{n'} - E_m}\Delta K + \cdots.$$

We omit number K everywhere on the right side because it is superfluous to indicate that mechanical anharmonicity parameters a_p and frequency ω are somewhat modified when free rotation is taken into account; these alterations are especially significant when K has large values. Moreover, if all calculations are performed in the second order of perturbation theory ($\sim \lambda^2$), then according to

$$(n|\rho_r|n') \sim \lambda^3, \quad \text{in which } n \neq n',$$

we can neglect the dependence of a_p and ω on K.

Raising quantity $(nK|d|n'K')$ to the second power, we obtain the algebraic expression for the first Herman–Wallis coefficient in the following simple form,

$$C_{nn'} = \frac{2\hbar\omega}{(n|d|n')} \sum_{m\neq n'} (n|d|m)\frac{(m|\rho_r|n')}{E_{n'} - E_m},$$

in which E_n, $(n|\rho_r|n')$ and $(n|d|n')$ are determined with a sufficient accuracy by convolutions of polynomials of quantum numbers [18–20]. The principal advantage of the formula obtained is that it is an expansion in terms of vibrational matrix elements $(n|d|m)$ that are diagonal in extraneous quantum number J (or K), which allows one to control the accuracy and the order of calculations.

To define the next coefficient, we calculate the second derivative of vector $|nK)$ with respect to K and the corresponding matrix element in the expansion of $(nK|d|n'K')$:

$$\begin{aligned}(nK|d\,\partial^2/\partial K^2|n'K) &= (\hbar\omega)^2\sum_{m\neq n'}\sum_{p\neq m}(n|d|p)\frac{(p|\rho_r|m)(m|\rho_r|n')}{(E_{n'}-E_m)(E_m-E_p)}\\ &+(\hbar\omega)^2\sum_{m\neq n'}\sum_{p\neq m}(n|d|m)\frac{(m|\rho_r|p)(p|\rho_r|n')}{(E_{n'}-E_m)(E_m-E_p)}\\ &+(\hbar\omega)^2\sum_{m\neq n'}\sum_{p\neq n'}(n|d|m)\frac{(m|\rho_r|p)(p|\rho_r|n')}{(E_{n'}-E_m)(E_{n'}-E_p)}\\ &-\hbar\omega\sum_{m\neq n'}\frac{E'_{n'}-E'_m}{(E_{n'}-E_m)^2}(n|d|m)(m|\rho_r|n').\end{aligned}$$

For simplicity, we omit number K here; one should bear in mind that E'_n denotes the first derivative of energy E_n with respect to K. Then, raising $(nK|d|n'K')$ to the second power, we obtain $D_{nn'}$; known contribution $C^2_{nn'}/4$ appears in the final expression, namely,

$$D_{nn'}=\frac{C^2_{nn'}}{4}+\frac{(nK|d\,\partial^2/\partial K^2|n'K)}{(n|d|n')}.$$

An arbitrary ℓth coefficient of the Herman–Wallis factor requires knowledge of derivative $\partial^\ell/\partial K^\ell|nK)$ and is trivially related to the preceding coefficients of the same factor.

The First Coefficients

We consider the calculation of the first coefficients $C_{0n'}$ up to $n'=3$ with the HI molecule as an example. An increased interest in HI is not fortuitous from the viewpoint of electro-optics. First, HI possesses an exceptional dipolar-moment function, which decreases with increasing bond length [53]:

$$d(q)=0.4471-0.0770\cdot(\lambda q)+0.547\cdot(\lambda q)^2-1.93\cdot(\lambda q)^3.$$

Second, through a small first derivative d', the intensity of the fundamental transition is small. For this reason, the first Herman–Wallis coefficient C_{01}(HI) is large, ~ 10 times that of corresponding values for other hydrogen halides [50,56]. Third, the dipolar-moment matrix elements for this molecule with perturbation theory should be constructed carefully because this molecule is characterized with strong electro-optical anharmonicity. We show above (see section 'Electro-Optics of Molecules') that the purely vibrational matrix elements of the HI dipolar moment, including the third and fourth overtones, are well reproduced by the polynomial perturbation theory in second order.

According to the alternative method, for vibrational dipolar-moment matrix elements of HI, one might use their experimental values/debye [47]:

$$(0|d|0)\approx d^0=0.4471,(0|d|1)=-0.00407,(0|d|2)=0.00197,(0|d|3)=-0.00113.$$

We perform all calculations without transcending the second order. This condition applies equally to the matrix elements of function ρ_r, which is a rigorous expansion in λ:

$$(n|\rho_r|n+1)=\sqrt{\frac{g_{n,n+1}}{2}}\left(\rho_r'+\left(\frac{1}{4}(11a^2-6A)\rho_r'-\frac{5}{2}a\rho_r''+\frac{1}{4}\rho_r'''\right)(n+1)\right),$$

$$(n|\rho_r|n+2)=\frac{1}{4}\sqrt{g_{n,n+2}}(2a\rho_r'+\rho_r''),$$

$$(n|\rho_r|n+3)=\frac{1}{4}\sqrt{\frac{g_{n,n+3}}{2}}\left(\left(\frac{3}{2}a^2+A\right)\rho_r'+2a\rho_r''+\frac{1}{3}\rho_r'''\right).$$

These matrix elements follow trivially from the general expression,

$$(n|\rho_r|n+k)=\sqrt{g_{n,n+k}}\sum_{s\alpha}\frac{2^{-s/2}}{s!}\rho_r^{(s)}\Pi^s_{(\beta\gamma)\alpha}(k),\quad s+\alpha\leq 3.$$

Here, $\lambda=1$ because

$$\rho_r^{(s)}=(-1)^s(s+1)!\lambda^{s+2}.$$

For HI, $\lambda=\sqrt{2B_e/\nu_e}=0.0751$ [52]; hence,

$$\rho_r'=-2\lambda^3=-8.5\cdot 10^{-4},\quad \rho_r''=6\lambda^4=2\cdot 10^{-4},\quad \rho_r'''=-24\lambda^5=-0.6\cdot 10^{-4}.$$

For HI, we present the required values of force parameters [49]:

$$a=-0.095\ \text{and}\ A=0.0103,$$

and proceed to the calculations.

The first coefficient C_{01} is simplest; it is practically determined in the harmonic approximation: $E_{n'}-E_m\approx\hbar\omega(n'-m)$, $a=0$, $A=0$ and $\rho_r^{(s)}=0$ for $s>1$, so that

$$C_{01}=\frac{2}{(0|d|1)}\sum_{m\neq 1}(0|d|m)\frac{(m|\rho_r|1)}{1-m}=\frac{2d^0}{(0|d|1)}\frac{\rho_r'}{\sqrt{2}}.$$

In the case of HI,

$$C_{01}(\mathrm{HI}) = 0.132(0.132).$$

We emphasize satisfactory agreement with experimental value (shown in parentheses) and those previously calculated [47].

We consider further the expression for coefficient C_{02}:

$$C_{02} = \frac{2\hbar\omega}{(0|d|2)}\left(\frac{(0|d|0)}{E_2 - E_0}(0|\rho_r|2) + \frac{(0|d|1)}{E_2 - E_1}(1|\rho_r|2) + \frac{(0|d|3)}{E_2 - E_3}(3|\rho_r|2)\right).$$

The largest contribution to both C_{01} and C_{02} of HI is made by permanent dipolar moment d^0, the magnitude of which is at least 100 times that of matrix elements $(0|d|k)$ with $k \neq 0$. As a result, retaining the first-order mechanical anharmonicity $(a \neq 0,\ a^2 = A = 0)$ with a zero-order approximation, we obtain

$$C_{02}(\mathrm{HI}) = \frac{2}{(0|d|2)}\left(\frac{d^0}{2}(0|\rho_r|2) + (0|d|1)(1|\rho_r|2) - (0|d|3)(3|\rho_r|2)\right) = 3.10 \cdot 10^{-2}.$$

The experimental value is $C_{02}(\mathrm{HI}) = 3.10 \cdot 10^{-2}$ [47].

In the case of C_{03}, one must perform all calculations carefully in the same order because even

$$E_{n'} - E_m \neq \hbar\omega(n' - m) \text{ and } (0|d|0) \neq d^0.$$

Using the following consideration, we estimate C_{03} with satisfactory accuracy in the adopted approximation; by definition, we have

$$C_{03} = \frac{2}{(0|d|3)}\left(\frac{(0|d|0)}{E_3 - E_0}(0|\rho_r|3) + \frac{(0|d|1)}{E_3 - E_1}(1|\rho_r|3) + \frac{(0|d|2)}{E_3 - E_2}(2|\rho_r|3) + \cdots\right).$$

One must calculate matrix element $(0|\rho_r|3)$ in the second order of mechanical anharmonicity and $(1|\rho_r|3)$ in the first order, whereas a value of $(2|\rho_r|3)$ should be evaluated in the harmonic approximation. Then,

$$C_{03}(\mathrm{HI}) \approx \frac{2}{(0|d|3)}\left(\frac{d^0}{3}(0|\rho_r|3) + \frac{(0|d|1)}{2}(1|\rho_r|3) + \frac{(0|d|2)}{1}\sqrt{\frac{3}{2}}\rho_r'\right) = 1.34 \cdot 10^{-2}.$$

This result is near a value that would be obtained with an inclusion of higher-order contributions involving $(0|d|4)$ and $(0|d|5)$ and agrees satisfactorily with experimental value $1.22 \cdot 10^{-2}$ [53].

Calculation of Higher-Order Approximations

Despite the fact that the first coefficients are obtainable easily and with a superior accuracy, higher-order approximations are equally important because theory and experiment cannot be properly compared without them. The Herman–Wallis coefficients are determined from experimentally evaluated derivatives or matrix elements of dipolar moment. An exact algorithm for calculating these coefficients is thus considered as a test of electro-optical anharmonicity.

We consider second coefficient $D_{nn'}$, which has a form

$$D_{nn'} = \frac{C_{nn'}^2}{4} + \frac{(nK|d\partial^2/\partial K^2|n'K)}{(n|d|n')}.$$

It is necessary to evaluate the matrix element of quantity $d\,\partial^2/\partial K^2$, the expression for which is represented as an expansion in terms of vibrational matrix elements $(n|d|n+k)$ and is quadratic in function ρ_r. Assuming that $E_{n'} - E_m \approx \hbar\omega(n'-m)$, we write this element as

$$\begin{aligned}
&(nK|d\,\partial^2/\partial K^2|n'K) = \\
&(n|d|n)\left[\sum_{p\neq n,n'}\frac{(n|\rho_r|p)(p|\rho_r|n')}{(n'-p)(p-n)} + \sum_{p\neq n}\frac{(n|\rho_r|p)(p|\rho_r|n')}{(n'-n)(n-p)} + \sum_{p\neq n'}\frac{(n|\rho_r|p)(p|\rho_r|n')}{(n'-n)(n'-p)}\right] \\
&+ (n|d|n+1)\left[\sum_{p\neq n+1,n'}\frac{(n+1|\rho_r|p)(p|\rho_r|n')}{(n'-p)(p-n-1)} + \sum_{p\neq n+1}\frac{(n+1|\rho_r|p)(p|\rho_r|n')}{(n'-n-1)(n+1-p)}\right. \\
&\qquad\left. + \sum_{p\neq n'}\frac{(n+1|\rho_r|p)(p|\rho_r|n')}{(n'-n-1)(n'-p)}\right] + \cdots.
\end{aligned}$$

Through the static dipolar moment, the main contribution appears from the first term containing $(n|d|n)$; the other expansion terms containing $(n|d|n+2)$, $(n|d|n+3)$, etc. decrease monotonically, thereby ensuring a uniform convergence of the perturbative expansion of matrix element $(nK|d\,\partial^2/\partial K^2|n'K)$. We deliberately omit the undoubtedly small contribution

$$-\hbar\omega\sum_{m\neq n'}\frac{E'_{n'}-E'_m}{(E_{n'}-E_m)^2}(n|d|m)(m|\rho_r|n'),$$

because, through the theorem of extraneous quantum numbers,

$$E'_{n'} - E'_m = (n'K|\rho_r|n'K) - (mK|\rho_r|mK) = -\hbar\omega\sigma(n'-m) + \cdots,$$

in which $\sigma \sim a\rho_r' \sim \lambda^4$.

We write coefficient D_{01} as

$$D_{01} = \frac{C_{01}^2}{4} - \frac{(0|d|0)}{(0|d|1)}\left[2(0|\rho_r|2)(1|\rho_r|2) + (0|\rho_r|3)(1|\rho_r|3) + \cdots\right] + \cdots.$$

If the value of matrix element $(1|\rho_r|2)$ is calculated in the harmonic approximation, product $(0|\rho_r|3)(1|\rho_r|3)$ and the following smaller contributions should be omitted, so that the equality

$$D_{01} = \frac{C_{01}^2}{4} - 2\rho_r'(0|\rho_r|2)\frac{(0|d|0)}{(0|d|1)}$$

is valid with sufficient accuracy. In the harmonic approximation, $C_{01}^2 \sim \rho_r'^2$, whereas $\rho_r'(0|\rho_r|2) \sim \rho_r'\rho_r''$; consequently, D_{01} should be approximately equal to $C_{01}^2/4$. For example, for HI,

$$D_{01}(\mathrm{HI}) = C_{01}^2/4 = 4.356 \cdot 10^{-3},$$

which agrees with experimental value $4.4 \cdot 10^{-3}$ [47].

The following coefficient D_{02} has a form

$$D_{02} = \frac{C_{02}^2}{4} + \frac{(0|d|0)}{(0|d|2)}\left[(0|\rho_r|1)(1|\rho_r|2) - (0|\rho_r|3)(2|\rho_r|3) + \cdots\right] + \cdots.$$

Like C_{01}, it can be calculated in the harmonic approximation; assuming that $(0|\rho_r|1)(1|\rho_r|2) \approx \rho_r'^2/\sqrt{2}$, we obtain

$$D_{02} = \frac{C_{02}^2}{4} + \frac{(0|d|0)}{(0|d|2)} \cdot \frac{\rho_r'^2}{\sqrt{2}}.$$

The value of coefficient C_{02} is of order $\rho_r''(0|d|0)/(0|d|2)$; because the absolute value of $(0|d|0)$ is at least 100 times that of $(0|d|2)$, coefficient D_{02} is thus primarily determined by quantity

$$\rho_r'^2(0|d|0)/\sqrt{2}(0|d|2),$$

and its sign is governed by the sign of ratio $(0|d|0)/(0|d|2)$. We emphasize that, for all hydrogen halides except HI, $(0|d|2)$ is negative and, because $(0|d|0) > 0$, we have $D_{02} < 0$. In contrast, for HI, $(0|d|2) > 0$ and D_{02} disagrees with experimental value $-2.5 \cdot 10^{-4}$ [47]; this coefficient is positive:

$$D_{02}(\mathrm{HI}) = 3.5 \cdot 10^{-4}.$$

Coefficient C_{01} is positive only for HI and only because matrix element $(0|d|1)$ is negative; for HF, HCl and HBr, $(0|d|1) > 0$ and $C_{01} < 0$ [50]. Taking into account that, from the viewpoint of calculations, D_{02} is similar to coefficient C_{01}, one readily concludes that the sign of D_{02} depends on the sign of $(0|d|2)$; D_{02} should hence be positive for HI and negative for other hydrogen halides.

We have the following expression for coefficient D_{03}:

$$D_{03} = \frac{C_{03}^2}{4} + \frac{(0|d|0)}{3(0|d|3)}\left[2(0|\rho_r|2)(2|\rho_r|3) + (0|\rho_r|1)(1|\rho_r|3) + \cdots\right] + \frac{(0|d|1)}{(0|d|3)}\left[(1|\rho_r|2)(2|\rho_r|3) + \cdots\right] + \cdots.$$

In the expansion of the coefficient in terms of purely vibrational matrix elements, we must retain here, with the main contribution containing $(0|d|0)$, the terms originating from $(0|d|1)$. Assuming that

$$(1|\rho_r|3) = \sqrt{3}(0|\rho_r|2)$$

and restricting ourselves to the first non-vanishing approximation, we readily verify that these contributions have almost the same order:

$$D_{03} = \frac{C_{03}^2}{4} + \sqrt{\frac{3}{2}}\rho_r'(0|\rho_r|2) \cdot \frac{(0|d|0)}{(0|d|3)} + \sqrt{\frac{3}{2}}\rho_r'^2 \cdot \frac{(0|d|1)}{(0|d|3)}.$$

For HI, this formula yields a value $1.0 \cdot 10^{-4}$, in satisfactory agreement with experimental result $2.9 \cdot 10^{-4}$ [53].

The calculation of higher-order approximations of the coefficients of the Herman–Wallis factor with tests for hydrogen halides is made elsewhere [50]; for HI, as an example, the corresponding results are listed in Table 3.2. Values [50] were obtained with the aid of the formalism of polynomials of quantum numbers. The matrix elements of function ρ_r were calculated in the fifth order:

$$(n|\rho_r|n+k) = \sqrt{g_{n,n+k}} \sum_{s\alpha} \frac{2^{-s/2}}{s!} \rho_r^{(s)} \Pi^s_{(\beta\gamma)\alpha}(k), \quad s + \alpha \le 5,$$

the quantities of energy

$$E_n = \hbar\omega\left(n + \frac{1}{2}\right) + \sum_{\alpha=1}^{4} \frac{\hbar\omega}{\alpha} \sum_{(p\nu)\alpha} p a_p \Pi^{p+2}_{(\beta\gamma)\nu}(0)$$

Table 3.2 Coefficients of the Herman–Wallis Factor for HI

Coefficient	Theory [50]	Theory [51]	Theory [47]	Experiment
$C_{01}\times 10^{1}$	1.32	1.32	1.32	1.32 [47]
$C_{02}\times 10^{2}$	3.09	3.17	3.14	3.10 [47]
$C_{03}\times 10^{2}$	1.25	1.16	1.26	1.22 [53]
$C_{04}\times 10^{2}$	1.45	1.63	1.41	1.76 [48]
$C_{05}\times 10^{2}$	1.73	1.69	1.86	1.73 [48]
$D_{01}\times 10^{3}$	4.35	3.50	3.60	4.40 [47]
$D_{02}\times 10^{4}$	3.62	2.50	− 3.60	− 2.50 [47]
$D_{03}\times 10^{4}$	0.98	0.23	− 0.14	2.90 [53]
$D_{04}\times 10^{4}$	1.08	0.66	1.70	20.5 [48]
$D_{05}\times 10^{4}$	1.45	2.40	2.90	15.4 [48]

in the fourth, and matrix elements of the dipolar-moment function were taken from Table 3.1 (see column Π-formalism).

This developed theory is readily extrapolated to further Herman–Wallis coefficients E, F, etc.; all coefficients clearly possess a definite structure. For example,

$$E_{nn'} = \frac{C_{nn'}}{2}\left(D_{nn'} - \frac{C_{nn'}^2}{4}\right) + \frac{(nK|d\,\partial^3/\partial K^3|n'K)}{3(n|d|n')}.$$

To evaluate the first two coefficients E_{01} and E_{02}, one might neglect the matrix element of quantity $d\,\partial^3/\partial K^3$. As a result, if we assume that $D_{01} = C_{01}^2/4$, we obtain $E_{01} = 0$, and

$$E_{02} = \frac{C_{02}}{2}\left(D_{02} - \frac{C_{02}^2}{4}\right).$$

To calculate E_{03}, one should evaluate matrix element $(0K|d\,\partial^3/\partial K^3|3K)$ in the harmonic approximation. This coefficient, from the viewpoint of calculations, is similar to quantities C_{01} and D_{02}; the higher-order corrections have no influence on the sign and order of these coefficients, which are defined mainly by the harmonic approximation. To evaluate the order of quantity E, note that $E \sim C \times D$, hence $E \sim 10^{-6}$. This estimate is rough but, as an order of magnitude, the value obtained reflects the real behaviour of the Herman–Wallis coefficients:

$$C \sim 10^{-2}, \quad D \sim 10^{-4}, \quad E \sim 10^{-6}, \ldots$$

Quantities E, F, etc. have little interest, through the formidable difficulties of their experimental determination, but from a theoretical point of view, higher-order coefficients are equal to zero only when all calculations are performed to low orders. In other cases, one must take them into account.

Future Developments

A prospective direction of further investigation is to find new solutions of the eigenvalue problems for vibration–rotational Hamiltonians represented in a convenient form to which our perturbation method becomes applicable. Everywhere above we use the wave functions of a harmonic oscillator as the basis functions, but the developed formalism is applicable successfully for basis functions of another choice. For instance, instead of a harmonic oscillator, one might use Morse's oscillator and, as a result, obtain another system of recurrence equations relating the matrix elements sought. In this respect, two pertinent remarks should be made.

First, one might find the energy of Morse's oscillator with the aid of perturbation theory, namely,

$$E_n = \hbar\omega\left(n+\frac{1}{2}\right)+\sum_{\alpha}\lambda^{\alpha}E_n^{\alpha},\quad E_n^{\alpha}=\frac{\hbar\omega}{\alpha}\sum_{(p\nu)\alpha}pa_p\Pi^{p+2}_{(\beta\gamma)\nu}(0).$$

The internuclear potential [57] with parameters D and a_M has this form,

$$V_M(x)=D(1-\mathrm{e}^{-a_M x})^2,$$

in which reduced variable $x=(r-r_0)/r_0$ characterizes the displacement of an oscillator of instantaneous internuclear separation r from its equilibrium value r_0. Expanding $V_M(x)$ in a power series in $x=\lambda q$, we find a_p:

$$a_p=\frac{(-a_M)^p(2^{p+1}-1)}{(p+2)!}2^{-\frac{(p+2)}{2}},\quad p>0.$$

Moreover, by definition, $\hbar\omega=2Da_M^2\lambda^2$. Clearly, $E_n^1=0$. For the second-order correction, we have

$$E_n^2=-\frac{\hbar\omega}{2}(7a_1^2-3a_2)-\hbar\omega(30a_1^2-6a_2)\left(n+\frac{1}{2}\right)^2=-\frac{(\hbar\omega)^2}{4D}\lambda^{-2}\left(n+\frac{1}{2}\right)^2.$$

The corrections beyond the second, after substitution of Morse's coefficients a_p into them, must equal zero. In this case only, the solution in perturbation theory coincides with exact solution [57] for energy,

$$E_n^{(M)} = \hbar\omega\left(n + \frac{1}{2}\right) - \frac{(\hbar\omega)^2}{4D}\left(n + \frac{1}{2}\right)^2.$$

This coincidence is genuine. Having calculated all pertinent polynomials, we are convinced that the corrections mentioned above are equal to zero, to at least the accuracy of the twentieth order. This result emphasizes once again the correctness of perturbation theory in the form of polynomials of quantum numbers.

To take into account an additional influence of anharmonicity, it suffices to modify the initial potential, i.e. to add terms of higher degrees of this quantity:

$$\xi_M = 1 - e^{-a_M x}.$$

For instance,

$$V'_M = V_M + \sum_{p>0} R_p \xi_M^{p+2},$$

in which appear coefficients R_p [6,58,59]; the extrapolation to the many-dimensional case is evidently not difficult. Quantity V_M is included in the zero-order Hamiltonian; other terms in the expression for a new potential V'_M should be considered with perturbation theory. For this purpose, one might use solution (3.9) and find matrix elements

$$\mathrm{M}^s_{\alpha\beta}(n, m) = \langle n,\ \alpha|\xi^s_M|m,\ \beta\rangle$$

of the perturbed Morse's oscillator. Corrections $|n,\alpha\rangle$ determine the exact wave function

$$|n) = |\psi_n\rangle + \sum_{\alpha>0} |n, \alpha\rangle.$$

To solve this problem, it suffices to calculate all necessary matrix elements of coordinate ξ_M to various degrees between wave functions ψ_n of the unperturbed Morse's oscillator. The correctness of the choice of these functions represents the second pertinent aspect.

For an oscillator of mass μ, functions ψ_n are defined in the Schrödinger equation

$$\frac{\partial^2\psi_n}{\partial x^2} + \frac{2\mu r_0^2}{\hbar^2}(E_n^{(M)} - V_M)\psi_n = 0.$$

Setting initially $y = \mathrm{e}^{-a_M x}$, then $z = 2Cy$ and $\psi_n = \mathrm{e}^{-z/2} z^{b/2} F_n(z)$, we obtain

$$z\frac{\partial^2 F_n}{\partial z^2} + (b + 1 - z)\frac{\partial F_n}{\partial z} + (C - b/2 - 1/2)F_n = 0,$$

in which

$$C = \frac{r_0}{\hbar a_M}\sqrt{2\mu D} \text{ and } b^2 = -4C^2\frac{E_n^{(M)} - D}{D}.$$

This equation has a well-known solution in the form of Laguerre polynomials (or confluent hypergeometric functions); namely,

$$F_n(z) = L_n^b(z) = \frac{\mathrm{e}^z}{z^b}\frac{\mathrm{d}^n}{\mathrm{d}z^n}(\mathrm{e}^{-z} z^{n+b})$$

under a condition that $C - b/2 - 1/2 = n$, in which $n = 0$, 1, 2, etc. This condition determines the possible values of energy $E_n^{(M)}$. In turn,

$$\psi_n = N_n\, \mathrm{e}^{-z/2} z^{C-n-1/2} L_n^{2C-2n-1}(z).$$

Normalization coefficient N_n is always chosen such that $\langle\psi_n|\psi_n\rangle = 1$. It seems that we have obtained the solution in an explicit form for use in calculating the matrix elements. This solution has meaning only when z belongs to an interval from 0 to $+\infty$, whereas for real molecules x has values from -1 to $+\infty$. The upper boundary value of z is hence equal to $2C\exp(a_M)$, and the zero-order functions become orthogonal only in a limit $C \to \infty$ [6]. This circumstance dampens the interest in Morse's basis functions to apply to the anharmonicity of molecular vibrations. In some cases, it is nevertheless convenient to use Morse's potential, sacrificing physical rigour and assuming approximately that $0 \le z < \infty$ [60]. This assumption yields small errors in the calculated matrix elements $\langle\psi_n|\xi_M^s|\psi_{n'}\rangle$.

Functions of Quantum Numbers

Another prospective direction for further investigation is to proceed beyond solutions with perturbation theory. We assume that the effective internuclear potential is a real function that is represented as an expansion in a power series in terms of the normal coordinates. In this case, the procedure of quantization, i.e. the calculation of matrix elements of an arbitrary coordinate function, taking into account the influence of anharmonicity, is reduced to the sum of polynomials multiplied by factor $\sqrt{g}$:

$$(n|f|n + k) = \sqrt{g_{n,n+k}}\sum_{\ell\alpha}\sum_{(s)\ell}\frac{2^{-\ell/2}}{\ell!}f_s^{(\ell)}\Pi^s_{(\beta\gamma)\alpha}(k).$$

For the anharmonic energy, we have a similar representation,

$$E_n = E_n^0 + \sum_{\alpha} \frac{1}{\alpha} \sum_{(p\nu)\alpha} p \sum_{(j)p+2} a_j \Pi^j_{(\beta\gamma)\nu}(0).$$

Expanding here the polynomials in terms of quantum numbers, we obtain this intriguing formula [16],

$$(n|f|n+k) = \sqrt{g_{n,n+k}} \sum_i \Phi^i_k \left(n + \frac{k+1}{2}\right)^i$$

for the one-dimensional case, and this one,

$$(n|f|n+k) = \sqrt{g_{n_1,n_1+k_1} \cdots g_{n_r,n_r+k_r}} \sum_{i_1 \cdots i_r} \Phi^{i_1 \cdots i_r}_{k_1 \cdots k_r} \left(n_1 + \frac{k_1+1}{2}\right)^{i_1} \cdots \left(n_r + \frac{k_r+1}{2}\right)^{i_r}$$

for the many-dimensional case. The validity of this expansion follows from the condition of the symmetry of the matrix element of function f, namely,

$$(n|f|n+k) = (n+k|f|n).$$

The substitution of number k here by $-k$ yields

$$(n|f|n-k) = (n-k|f|n) = \sqrt{g_{n-k,n}} \sum_i \Phi^i_{-k} \left(n + \frac{1-k}{2}\right)^i.$$

We obtain this expansion alternatively with the aid of a formal substitution of n by $n-k$, i.e. again,

$$(n|f|n-k) = (n-k|f|n) = \sqrt{g_{n-k,n}} \sum_i \Phi^i_k \left(n + \frac{1-k}{2}\right)^i.$$

Assuming $\Phi^i_{-k} = \Phi^i_k$, we ascertain that the expansion of $(n|f|n+k)$ in powers of $n + k/2 + 1/2$ is valid. Quantity $k/2$ ensures the symmetry of the matrix element, whereas factor 1/2 appears because of the commutation relation between destruction operator η and creation operator η^+, $[\eta,\eta^+] = 1$. Note that $\xi = \eta + \eta^+$. For the many-dimensional case, the validity of the expansion above is established in a similar manner.

The derived expansions in terms of quantum numbers hold for the matrix elements of an arbitrary physical function that is represented as an expansion in a power series in terms of creation and destruction operators. This consequence of perturbation-theory calculations is trivial. The values of energy E_n are also expressible from the formula for $(n|f|n+k)$ in which $f = H$. Assuming $k = 0$, we obtain

$$E_n = \sum_{i_1 i_2 \cdots i_r} \Omega_{i_1 i_2 \cdots i_r} \left(n_1 + \frac{1}{2}\right)^{i_1} \left(n_2 + \frac{1}{2}\right)^{i_2} \cdots \left(n_r + \frac{1}{2}\right)^{i_r},$$

in which mechanical anharmonicity parameters $\Omega_{i_1 i_2 \cdots i_r}$ might be expressed through a_j and ω_k. To generalize our theory, we assume that quantity E_n is a function of quantum numbers $n_1+1/2$, $n_2+1/2, \ldots$, $n_r+1/2$. Together with this dependence on the quantum numbers, energy E_n depends on some parameters that exhibit the influence of anharmonicity. Altering these parameters and an explicit form of E_n in various manners, we obtain various representations of anharmonicity.

We can determine heuristically a function Φ for the matrix element of a particular physical quantity $f(\xi)$, for instance, the dipolar moment, as a dependence on quantum number $n+k/2+1/2$:

$$(n|f|n+k) = \sqrt{g_{n,n+k}}\Phi_k\left(n + \frac{k+1}{2}\right).$$

Functions Φ_k are arbitrarily expressible here, e.g.

$$\Phi_k = \theta_k(n + k/2 + 1/2)e^{-\phi_k(n+k/2+1/2)}, \quad \Phi_k = \theta_k(n+k/2+1/2)^{-1}, \text{ etc.,}$$

with parameters θ_k and ϕ_k determined from experiment. In the present formalism, one might also construct phenomenologically a function $\Phi_{k_1 k_2 \ldots k_r}$ for a system with r variables. However, from the solution of the Schrödinger equation according to perturbation theory, follow not the functions themselves rather their expansions in terms of quantum numbers with coefficients Φ_k^i that characterize the exact influence of anharmonicity. These coefficients have no dependence on quantum numbers and have the dimension of initial function f. The introduction of the functions of quantum numbers is essentially a conversion to an 'anharmonicity representation', which transcends the solution according to perturbation theory. The study of these functions of quantum numbers with pertinent laws represents a special interest in physics today.

Background

First Polynomial Convolutions $\Pi^{s}_{(\beta\gamma)\alpha}(k)$

s	$\alpha = 0$	$\alpha = 1$
1	$\Pi^{1}_{(\beta\gamma)0}(1) = 1$	$\Pi^{1}_{(\beta\gamma)1}(0) = -6a_1(2n+1)$, $\Pi^{1}_{(\beta\gamma)1}(2) = 2a_1$
2	$\Pi^{2}_{(\beta\gamma)0}(0) = 2n+1$, $\Pi^{2}_{(\beta\gamma)0}(2) = 1$	$\Pi^{2}_{(\beta\gamma)1}(1) = -20a_1(n+1)$, $\Pi^{2}_{(\beta\gamma)1}(3) = 4a_1$
3	$\Pi^{3}_{(\beta\gamma)0}(1) = 3(n+1)$, $\Pi^{3}_{(\beta\gamma)0}(3) = 1$	$\Pi^{3}_{(\beta\gamma)1}(0) = -2a_1(30n^2+30n+11)$, $\Pi^{3}_{(\beta\gamma)1}(2) = -12a_1(2n+3)$, $\Pi^{3}_{(\beta\gamma)1}(4) = 6a_1$
4	$\Pi^{4}_{(\beta\gamma)0}(0) = 6n^2+6n+3$, $\Pi^{4}_{(\beta\gamma)0}(2) = 4n+6$, $\Pi^{4}_{(\beta\gamma)0}(4) = 1$	$\Pi^{4}_{(\beta\gamma)1}(1) = -8a_1(14n^2+28n+17)$, $\Pi^{4}_{(\beta\gamma)1}(3) = -24a_1(n+2)$, $\Pi^{4}_{(\beta\gamma)1}(5) = 8a_1$
5	$\Pi^{5}_{(\beta\gamma)0}(1) = 10n^2+20n+15$, $\Pi^{5}_{(\beta\gamma)0}(3) = 5n+10$, $\Pi^{5}_{(\beta\gamma)0}(5) = 1$	$\Pi^{5}_{(\beta\gamma)1}(0) = -10a_1(2n+1)(14n^2+14n+13)$, $\Pi^{5}_{(\beta\gamma)1}(2) = -10a_1(17n^2+51n+43)$, $\Pi^{5}_{(\beta\gamma)1}(4) = -10a_1(2n+5)$, $\Pi^{5}_{(\beta\gamma)1}(6) = 10a_1$
6	$\Pi^{6}_{(\beta\gamma)0}(0) = 20n^3+30n^2+40n+15$, $\Pi^{6}_{(\beta\gamma)0}(2) = 15n^2+45n+45$, $\Pi^{6}_{(\beta\gamma)0}(4) = 6n+15$, $\Pi^{6}_{(\beta\gamma)0}(6) = 1$	$\Pi^{6}_{(\beta\gamma)1}(1) = -60a_1(n+1)(9n^2+18n+19)$, $\Pi^{6}_{(\beta\gamma)1}(3) = -4a_1(57n^2+228n+245)$, $\Pi^{6}_{(\beta\gamma)1}(5) = -12a_1(n+3)$, $\Pi^{6}_{(\beta\gamma)1}(7) = 12a_1$
7	$\Pi^{7}_{(\beta\gamma)0}(1) = 35n^3+105n^2+175n+105$, $\Pi^{7}_{(\beta\gamma)0}(3) = 21n^2+84n+105$, $\Pi^{7}_{(\beta\gamma)0}(5) = 7n+21$, $\Pi^{7}_{(\beta\gamma)0}(7) = 1$	$\Pi^{7}_{(\beta\gamma)1}(0) = -210a_1(6n^4+12n^3+22n^2+16n+5)$, $\Pi^{7}_{(\beta\gamma)1}(2) = -112a_1(2n+3)(4n^2+12n+15)$, $\Pi^{7}_{(\beta\gamma)1}(4) = -140a_1(2n^2+10n+13)$, $\Pi^{7}_{(\beta\gamma)1}(6) = 0$, $\Pi^{7}_{(\beta\gamma)1}(8) = 14a_1$
8	$\Pi^{8}_{(\beta\gamma)0}(0) = 70n^4+140n^3+350n^2+280n+105$, $\Pi^{8}_{(\beta\gamma)0}(2) = 56n^3+252n^2+532n+420$, $\Pi^{8}_{(\beta\gamma)0}(4) = 28n^2+140n+210$, $\Pi^{8}_{(\beta\gamma)0}(6) = 8n+28$, $\Pi^{8}_{(\beta\gamma)0}(8) = 1$	$\Pi^{8}_{(\beta\gamma)1}(1) = -112a_1(22n^4+88n^3+202n^2+228n+105)$, $\Pi^{8}_{(\beta\gamma)1}(3) = -224a_1(n+2)(6n^2+24n+35)$, $\Pi^{8}_{(\beta\gamma)1}(5) = -32a_1(10n^2+60n+91)$, $\Pi^{8}_{(\beta\gamma)1}(7) = 16a_1(n+4)$, $\Pi^{8}_{(\beta\gamma)1}(9) = 16a_1$

(Continued)

Background (Continued)

First Polynomial Convolutions $\Pi^{s}_{(\beta\gamma)\alpha}(k)$

s	$\alpha = 2$
1	$\Pi^{1}_{(\beta\gamma)2}(1) = 2(11a_1^2 - 3a_2)(n+1), \quad \Pi^{1}_{(\beta\gamma)2}(3) = 3a_1^2 + a_2$
2	$\Pi^{2}_{(\beta\gamma)2}(0) = a_1^2(240n^2 + 240n + 88) - a_2(24n^2 + 24n + 12),$
	$\Pi^{2}_{(\beta\gamma)2}(2) = (a_1^2 - 5a_2)(2n+3), \quad \Pi^{2}_{(\beta\gamma)2}(4) = 10a_1^2 + 2a_2$
3	$\Pi^{3}_{(\beta\gamma)2}(1) = a_1^2(519n^2 + 1038n + 612) - a_2(51n^2 + 102n + 72),$
	$\Pi^{3}_{(\beta\gamma)2}(3) = -12(4a_1^2 + a_2)(n+2), \quad \Pi^{3}_{(\beta\gamma)2}(5) = 21a_1^2 + 3a_2$
4	$\Pi^{4}_{(\beta\gamma)2}(0) = 4(2n+1)(a_1^2(225n^2 + 225n + 171) - a_2(17n^2 + 17n + 21)),$
	$\Pi^{4}_{(\beta\gamma)2}(2) = a_1^2(748n^2 + 2244n + 1834) - a_2(84n^2 + 252n + 234),$
	$\Pi^{4}_{(\beta\gamma)2}(4) = -2(29a_1^2 + 3a_2)(2n+5), \quad \Pi^{4}_{(\beta\gamma)2}(6) = 36a_1^2 + 4a_2$
5	$\Pi^{5}_{(\beta\gamma)2}(1) = 5(n+1)(a_1^2(771n^2 + 1542n + 1378) - a_2(55n^2 + 110n + 138)),$
	$\Pi^{5}_{(\beta\gamma)2}(3) = a_1^2(840n^2 + 3360n + 3495) - a_2(120n^2 + 480n + 555),$
	$\Pi^{5}_{(\beta\gamma)2}(5) = -10(19a_1^2 + a_2)(n+3), \quad \Pi^{5}_{(\beta\gamma)2}(7) = 55a_1^2 + 5a_2$
6	$\Pi^{6}_{(\beta\gamma)2}(0) = a_1^2(10860n^4 + 21720n^3 + 33360n^2 + 22500n + 6460)$
	$- a_2(660n^4 + 1320n^3 + 2760n^2 + 2100n + 720),$
	$\Pi^{6}_{(\beta\gamma)2}(2) = 3(2n+3)(a_1^2(1083n^2 + 3249n + 3515) - a_2(79n^2 + 237n + 335)),$
	$\Pi^{6}_{(\beta\gamma)2}(4) = a_1^2(732n^2 + 3660n + 4560) - a_2(156n^2 + 780n + 1080),$
	$\Pi^{6}_{(\beta\gamma)2}(6) = -3(43a_1^2 + a_2)(2n+7), \quad \Pi^{6}_{(\beta\gamma)2}(8) = 78a_1^2 + 6a_2$
7	$\Pi^{7}_{(\beta\gamma)2}(1) = a_1^2(23135n^4 + 92540n^3 + 191730n^2 + 198380n + 83860)$
	$- a_2(1323n^4 + 5292n^3 + 12978n^2 + 15372n + 7560),$
	$\Pi^{7}_{(\beta\gamma)2}(3) = 21(n+2)(a_1^2(447n^2 + 1788n + 2285) - a_2(35n^2 + 140n + 225)),$
	$\Pi^{7}_{(\beta\gamma)2}(5) = a_1^2(385n^2 + 2310n + 3143) - a_2(189n^2 + 1134n + 1827),$
	$\Pi^{7}_{(\beta\gamma)2}(7) = -308a_1^2(n+4), \quad \Pi^{7}_{(\beta\gamma)2}(9) = 105a_1^2 + 7a_2$

8	$\Pi^{8}_{(\beta\gamma)2}(0) = 56(2n+1)(a_1^2(531n^4 + 1062n^3 + 2794n^2 + 2263n + 1325) - a_2(27n^4 + 54n^3 + 198n^2 + 171n + 135))$,
	$\Pi^{8}_{(\beta\gamma)2}(2) = a_1^2(41552n^4 + 249312n^3 + 682220n^2 + 924756n + 506380) - a_2(2352n^4 + 14112n^3 + 43260n^2 + 66276n + 41580)$,
	$\Pi^{8}_{(\beta\gamma)2}(4) = 4(2n+5)(a_1^2(1508n^2 + 7540n + 11081) - a_2(132n^2 + 660n + 1197))$,
	$\Pi^{8}_{(\beta\gamma)2}(6) = -a_1^2(216n^2 + 1512n + 3388) - a_2(216n^2 + 1512n + 2772)$,
	$\Pi^{8}_{(\beta\gamma)2}(8) = -4(41a_1^2 - a_2)(2n+9)$, $\Pi^{8}_{(\beta\gamma)2}(10) = 136a_1^2 + 8a_2$

	$E_n = \hbar\omega\sum_i X_i(n+1/2)^i$
i	Anharmonicity parameters X_i
1	$1 - 1155a_1^4 + 918a_1^2a_2 - 67a_2^2 - 190a_1a_3 + 25a_4$
2	$-30a_1^2 + 6a_2 - 418110a_1^6 + 479970a_1^4a_2 - 124026a_1^2a_2^2 - 95460a_1^3a_3 + 29340a_1a_2a_3 + 3414a_2^3 + 17070a_1^2a_4 - 2730a_1a_5 - 1085a_3^2 - 1770a_2a_4 + 245a_6$
3	$-2820a_1^4 + 1800a_1^2a_2 - 68a_2^2 - 280a_1a_3 + 20a_4$
4	$-463020a_1^6 + 465300a_1^4a_2 - 99780a_1^2a_2^2 - 78120a_1^3a_3 + 19320a_1a_2a_3 + 1500a_2^3 + 10860a_1^2a_4 - 1260a_1a_5 - 630a_3^2 - 660a_2a_4 + 70a_6$

Polynomial convolutions $\Pi^{s}_{(\beta\gamma)\alpha}(k)$, in which $\alpha = 0$, 1 and 2, and $s = 1, 2, \ldots, 8$, are listed in the table. The first convolutions were used to calculate the matrix elements of the dipolar-moment function for diatomic molecules in the second order (see section 'Electro-Optics of Molecules'). In this table, we also show anharmonicity parameters X_i that are obtained with the aid of the formalism of polynomials of quantum numbers in the sixth order.

In an analogous manner, one might evaluate the polynomial convolutions for a two-dimensional case, a three-dimensional case, etc. For instance,

$$\Pi^{1}_{(\beta\gamma)1}(0) = -12a_1(n + 1/2)$$

for one variable,

$$\Pi^{11}_{(\beta\gamma)1}(1,0) = -\frac{2}{\omega_2(4\omega_1^2 - \omega_2^2)}(a_{21}(8\omega_1^2 - 3\omega_2^2)(n_1 + 1) + (24a_{03}\omega_1^2 + 4a_{21}\omega_1\omega_2 - 6a_{03}\omega_2^2)(n_2 + 1/2))$$

for two variables, and

$$\Pi^{111}_{(\beta\gamma)1}(1,1,0) = -\frac{2}{\omega_3(4\omega_1^2 - \omega_3^2)(4\omega_2^2 - \omega_3^2)}(\pi_1(n_1 + 1) + \pi_2(n_2 + 1) + \pi_3(n_3 + 1/2))$$

for three, in which

$$\pi_1 = a_{201}(32\omega_1^2\omega_2^2 + 3\omega_3^4 - 8\omega_1^2\omega_3^2 - 12\omega_2^2\omega_3^2),$$
$$\pi_2 = a_{021}(32\omega_1^2\omega_2^2 + 3\omega_3^4 - 8\omega_2^2\omega_3^2 - 12\omega_1^2\omega_3^2),$$
$$\pi_3 = a_{003}(96\omega_1^2\omega_2^2 + 6\omega_3^4 - 24\omega_1^2\omega_3^2 - 24\omega_2^2\omega_3^2) + a_{201}(16\omega_1\omega_2^2\omega_3 - 4\omega_1\omega_3^3) + a_{021}(16\omega_1^2\omega_2\omega_3 - 4\omega_2\omega_3^3).$$

This simple example demonstrates how the extension of the number of variables complicates an explicit form of the coefficients in the polynomial convolution expansion in terms of quantum numbers $n_i + k_i/2 + 1/2$.

4 Effects of Anharmonicity

Extension to Magnetic Phenomena

In the quantum theory of magnetism, the method of second quantization based on the simple model of a harmonic oscillator is widely used to describe the spin interaction. One might imagine that the spin momenta, precessing around a chosen direction, vibrate, and that the elementary spin interaction can be taken into account with the aid of anharmonic corrections. Can we extend the ideas of optical anharmonicity to explain the physical effects of the non-linear nature of the spin interaction? Can we discover a non-linear dependence of the magnetic-dipolar moment on the spin variables? If answers to these questions are affirmative, in calculating the matrix elements in magneto-optics according to perturbation theory, the terms of the spin Hamiltonian compete with terms of the same order in the expansion of the magnetic-dipolar moment of the system in the spin variables [18,20].

We consider a system of spin momenta $\mathbf{S}_k(S_k^x, S_k^y, S_k^z)$, $k = 1, 2, \ldots, r$, with Hamiltonian H_s. Let the spins in the ground state be fully ordered, i.e. parallel and oriented along the z-axis. In this case, quantities S_k^x, S_k^y and $S_k - S_k^z$, in which S_k is a value of the total kth spin, are infinitesimal. Our aim is to evaluate the energies of states that are weakly excited when the ground state is determined by the minimum of the Hamiltonian under an additional condition $S_k - S_k^z = 0$. Having set that $\delta H_s/\delta S_k^\sigma = 0$, in which $\sigma = x$ or y, and $\delta H_s/\delta S_k^z = -\gamma_k$, we expand H_s in powers of the spin variables, restricting this expansion to the second order:

$$H_s^0 = \sum_k \gamma_k (S_k - S_k^z) + \sum_{kj} \sum_{\rho\sigma} \Gamma_{kj}^{\rho\sigma} S_k^\rho S_j^\sigma .$$

Quantity $\gamma_k(S_k - S_k^z)$ is here the Zeeman energy of the kth spin and coefficients $\Gamma_{kj}^{\rho\sigma}$ describe the spin interaction; $\Gamma_{kj}^{\rho\sigma} = \Gamma_{jk}^{\rho\sigma} = \Gamma_{kj}^{\sigma\rho} = (\Gamma_{kj}^{\rho\sigma})^*$.

If the spin alterations are small, one might conventionally proceed to apply the Bose operators for creation τ_k^+ and destruction τ_k,

$$S_k^+ = S_k^x + \mathrm{i}S_k^y = \sqrt{2S_k}\tau_k, \quad S_k^- = S_k^x - \mathrm{i}S_k^y = \sqrt{2S_k}\tau_k^+, \quad S_k^z = S_k - \tau_k^+\tau_k,$$

in which $[\tau_k, \tau_j^+] = \hbar\delta_{kj}$ and $[\tau_k, \tau_j] = 0$. Introducing the notations

$$\gamma_{kj} = \frac{\gamma_k}{2}\delta_{kj} + \sqrt{S_k S_j} \sum_{\rho\sigma} \Gamma_{kj}^{\rho\sigma} ((c_k^\rho)^* c_j^\sigma + (c_k^\sigma)^* c_j^\rho) = (\gamma_{jk})^*$$

Quantum Theory of Anharmonic Effects in Molecules. DOI: http://dx.doi.org/10.1016/B978-0-12-397912-4.00004-0

and

$$\varepsilon_{kj} = 2\sqrt{S_k S_j}\sum_{\rho\sigma}\Gamma_{kj}^{\rho\sigma}c_k^\rho c_j^\sigma = \varepsilon_{jk}, \quad c_k^x = \mathrm{i}c_k^y = 1/2,$$

we rewrite the Hamiltonian in a quadratic form with τ_k and τ_k^+:

$$H_s^0 = 2\sum_{kj}\gamma_{kj}\tau_k^+\tau_j + \sum_{kj}\varepsilon_{kj}\tau_k^+\tau_j^+ + \sum_{kj}\varepsilon_{kj}^*\tau_k\tau_j.$$

This substitution of variables is an approximate procedure commonly called the approximate second-quantization method. Its main idea consists of reducing H_s^0 further to a diagonal form, thus yielding the energy levels of the states weakly excited. We effect this operation through a convenient canonical transformation,

$$\tau_k = \sum_\alpha (A_{k\alpha}\eta_\alpha + B_{k\alpha}^*\eta_\alpha^+) \text{ and } \tau_k^+ = \sum_\alpha (A_{k\alpha}^*\eta_\alpha^+ + B_{k\alpha}\eta_\alpha),$$

in which coefficients $A_{k\alpha}$ and $B_{k\alpha}$ satisfy trivial conditions,

$$\sum_\alpha (A_{k\alpha}B_{j\alpha}^* - A_{j\alpha}B_{k\alpha}^*) = 0 \text{ and } \sum_\alpha (A_{k\alpha}A_{j\alpha}^* - B_{k\alpha}^*B_{j\alpha}) = \delta_{kj},$$

that follow from commutation relations $[\tau_k,\tau_j] = 0$, $[\tau_k, \tau_j^+] = \hbar\delta_{kj}$ and $[\eta_\alpha,\eta_\beta] = 0$, $[\eta_\alpha, \eta_\beta^+] = \hbar\delta_{\alpha\beta}$. There clearly exists an inverse transformation

$$\eta_\alpha = \sum_k (A_{k\alpha}^*\tau_k - B_{k\alpha}^*\tau_k^+) \text{ and } \eta_\alpha^+ = \sum_k (A_{k\alpha}\tau_k^+ - B_{k\alpha}\tau_k)$$

with additional conditions

$$\sum_k (A_{k\alpha}B_{k\beta} - A_{k\beta}B_{k\alpha}) = 0 \text{ and } \sum_k (A_{k\alpha}^*A_{k\beta} - B_{k\alpha}^*B_{k\beta}) = \delta_{\alpha\beta}. \tag{4.1}$$

The reduced Hamiltonian thus assumes a form

$$\begin{aligned} H_s^0 = & \textstyle\sum_{\alpha\beta}\eta_\alpha^+\eta_\beta\sum_{kj}\{(\gamma_{kj}A_{j\beta} + \varepsilon_{kj}B_{j\beta})A_{k\alpha}^* + (\gamma_{kj}A_{k\alpha}^* + \varepsilon_{kj}^*B_{k\alpha}^*)A_{j\beta}\} \\ & + \textstyle\sum_{\alpha\beta}\eta_\alpha\eta_\beta^+\sum_{kj}\{(\gamma_{kj}B_{j\beta}^* + \varepsilon_{kj}A_{j\beta}^*)B_{k\alpha} + (\gamma_{kj}B_{k\alpha} + \varepsilon_{kj}^*A_{k\alpha})B_{j\beta}^*\} \\ & + \textstyle\sum_{\alpha\beta}\eta_\alpha^+\eta_\beta^+\sum_{kj}\{(\gamma_{kj}B_{j\beta}^* + \varepsilon_{kj}A_{j\beta}^*)A_{k\alpha}^* + (\gamma_{kj}A_{k\alpha}^* + \varepsilon_{kj}^*B_{k\alpha}^*)B_{j\beta}^*\} \\ & + \textstyle\sum_{\alpha\beta}\eta_\alpha\eta_\beta\sum_{kj}\{(\gamma_{kj}A_{j\beta} + \varepsilon_{kj}B_{j\beta})B_{k\alpha} + (\gamma_{kj}B_{k\alpha} + \varepsilon_{kj}^*A_{k\alpha})A_{j\beta}\}. \end{aligned}$$

One must here choose $A_{k\alpha}$ and $B_{k\alpha}$ so that, in the Hamiltonian, only diagonal terms are retained, i.e. terms of type $\eta_\alpha^+\eta_\alpha$. This step is readily effected (see Ref. [61]), assuming that $A_{k\alpha}$ and $B_{k\alpha}$ are eigenfunctions of equations in the following system, with eigenvalues ω_α:

$$\sum_j (\gamma_{kj}A_{j\alpha} + \varepsilon_{kj}B_{j\alpha}) = \frac{1}{2}\omega_\alpha A_{k\alpha} \text{ and } \sum_j (\gamma_{kj}^* B_{j\alpha} + \varepsilon_{kj}^* A_{j\alpha}) = -\frac{1}{2}\omega_\alpha B_{k\alpha}. \quad (4.2)$$

Using Eq. (4.2) and taking into account that through Eq. (4.1)

$$\sum_{\alpha\beta} \eta_\alpha^+ \eta_\beta^+ (\omega_\alpha - \omega_\beta) \sum_k A_{k\alpha}^* B_{k\beta}^* = 0 \text{ and } \sum_{\alpha\beta} \eta_\alpha \eta_\beta (\omega_\alpha - \omega_\beta) \sum_k A_{k\alpha} B_{k\beta} = 0,$$

we obtain the expression sought:

$$H_s^0 = \sum_\alpha \omega_\alpha \eta_\alpha^+ \eta_\alpha - \hbar \sum_{\alpha k} \omega_\alpha B_{k\alpha}^* B_{k\alpha}.$$

The eigenvalues of H_s^0 are equal to

$$E_{sn}^0 = \hbar \sum_\alpha \omega_\alpha n_\alpha - \hbar \sum_{\alpha k} \omega_\alpha B_{k\alpha}^* B_{k\alpha}, \quad n_\alpha = 0, 1, 2, \ldots.$$

Energy E_{sn}^0 of the system of weakly interacting spins is thus determined by the harmonic Hamiltonian of a non-interacting elementary excitation – spin waves. Quantities η_α^+ and η_α are hence operators of creation and destruction of one spin wave with wave vector α. The principal condition of the validity of applying this approximate second-quantization method is that spin alterations $S_k - S_k^z$ are sufficiently small. The expectation values of the occupation numbers are consequently too small: $\langle n_\alpha \rangle \ll 1$. This description is valid only while we consider the states to be weakly excited.

Magneto-Optical Anharmonicity

We consider function μ for the magnetic-dipolar moment of a system. At first glance, the quantity of magnetic moment is proportional to S, i.e. increment of spin δS. According to a quasi-classical interpretation, the spin momentum precesses about z-axis so that its increment is virtually determined by variables S^+ and S^-. From a quantum-mechanical point of view, operators S^+ and S^-, acting on a state of the system, alter, in a first approximation, the occupation numbers of elementary excitations by unity. In a harmonic approximation, the spin increment, to be denoted by ξ, is therefore linear in variables τ and τ^+:

$$\xi = (\delta T)\tau + (\delta T)^* \tau^+.$$

Coefficient δT, in turn, is chosen such that the zero-order Hamiltonian H_s^0 is diagonal in the canonically conjugate Hermitian variables. As η and η^+ are such variables, we postulate an equality $\xi = \eta + \eta^+$, i.e.

$$\xi_\alpha = \eta_\alpha + \eta_\alpha^+ = \sum_k \{(A_{k\alpha}^* - B_{k\alpha})\tau_k + (A_{k\alpha}^* - B_{k\alpha})^* \tau_k^+\}.$$

Because $\mu \sim \xi$, in the harmonic approximation, the magnetic-dipolar-moment function of the system is linear in the spin–wave variables.

Considering the quadratic form of the spin Hamiltonian, we ascertain that the system of weakly interacting spins is converted into the system of non-interacting spin waves. In this case, the magnetic-dipolar moment is simply linear in the spin-wave variables. We should clearly consider spin–spin interaction W_s beyond the second order in quantities τ_k or ξ_α. For this purpose, we rewrite the Hamiltonian in a form

$$H_s = H_s^0 + W_s,$$

in which function $W_s(\xi_1, \xi_2, \ldots)$ that characterizes the energy of the spin–spin interaction is expressible as a power series in variables ξ_α:

$$W_s = \sum_{p>0} \delta^p \sum_{(j_1 j_2 \cdots j_r)p+2} b_{j_1 j_2 \cdots j_r} \xi_1^{j_1} \xi_2^{j_2} \cdots \xi_r^{j_r};$$

this expansion leads us, in essence, to a concept of spin–wave anharmonicity. Here, b_j are spin anharmonic parameters similar to a_j; δ, like λ, specifies the order of the spin–wave interaction.

In this form, the problem of the eigenvalues E_{sn} of Hamiltonian H_s has a natural solution in the form of polynomials of quantum numbers. The values of differences of quantities E_{sn} determine the possible transition frequencies of the spin system. Not all transitions manifest themselves in an experiment: some are improbable (magnetic resonance). The selection rules for these resonance transitions are governed by the matrix elements of magnetic-dipolar moment μ. Because the correct Hamiltonian incorporating the anharmonicity contains spin operators of higher orders, the part of μ non-linear in the spin variables can also significantly affect the transition intensities (non-linear magnetic resonance). This effect is mathematically encompassed in terms of the magneto-optical anharmonicity reflecting the non-linear nature of the magnetic-dipolar-moment function,

$$\mu = \sum_\ell \sum_{(s)\ell} \frac{2^{-\ell/2}}{\ell!} \mu_s^{(\ell)} \xi_1^{s_1} \xi_2^{s_2} \cdots \xi_r^{s_r},$$

in which $\mu_s^{(\ell)}$ are derivatives of the magnetic-dipolar moment. One can see that this situation is completely analogous to the electro-optics of molecules. It would be inconsistent to incorporate anharmonicity into only Hamiltonian H_s but to consider function μ to be linear in the spin variables.

The magneto-dynamical effect is insignificant. To estimate this effect, we set $\Gamma_{kj}^{\rho\sigma} = \Gamma$, $\gamma_k = \gamma$, $\varepsilon_{kj} = 0$, $A_{k\alpha} = r^{-1/2}\,\mathrm{e}^{-\mathrm{i}\alpha\mathbf{x}_k}$ and $B_{k\alpha} = 0$, that is appropriate for an isotropic ferromagnetic system, and through Eq. (4.2) obtain

$$\omega_\alpha = 2\sum_j \gamma_{kj}\exp(\mathrm{i}\alpha\mathbf{x}_{kj}) = \gamma + 2\Gamma\sum_j \sqrt{S_k S_j}\exp(\mathrm{i}\alpha\mathbf{x}_{kj}),$$

in which $\mathbf{x}_k$ is the radius vector of the kth spin; $\mathbf{x}_{kj} = \mathbf{x}_k - \mathbf{x}_j$. Restricting to l near neighbours in a summation over j and assuming $S_k = S$, we have

$$\omega_\alpha = \gamma + 2\Gamma S\sum_j \exp(\mathrm{i}\alpha\mathbf{x}_{kj}) \approx \gamma + 2\Gamma S l - \Gamma S\sum\nolimits_j(\alpha\mathbf{x}_{kj})^2.$$

Quantity $2\Gamma S l$ is exactly compensated by the same contribution that appears in γ with the opposite sign. Recall that γ is defined from the condition of the minimum of the ground-state energy. The part of γ that is non-compensated determines the Zeeman energy. Whereas the latter is a few cm^{-1} of order, the typical magnon energy,

$$\Gamma S\sum_j (\alpha\mathbf{x}_{kj})^2 \sim \Gamma S(a\alpha)^2,$$

amounts only to 10^{-1}–$10^{-3}\,\mathrm{cm}^{-1}$, in which $|a\alpha| \sim 10^{-2}$–10^{-3} [32], which is valid for magnons with microwave frequencies. In this estimation, we take into account the value of exchange interaction energy $\Gamma \sim e^2/a \sim 10^3\,\mathrm{cm}^{-1}$, in which e is the absolute charge of the electron and a is the lattice parameter, typically about 2–3 Å. As we see, the anharmonic corrections amount to 10^{-3}–$10^{-5}\,\mathrm{cm}^{-1}$ at most and, from just this point, the effect of magneto-optical anharmonicity on the matrix elements of magnetic moment μ becomes important.

An Electron in a Magnetic Field

The Dirac equation supports exact solutions for a freely moving electron and for the case of Coulomb potential, but other exact solutions exist. For instance, one might readily obtain energy levels for an electron in a homogeneous magnetic field. According to the non-relativistic Schrödinger theory, we have a similar structure of the energy levels, which are known as Landau levels.

Understanding the specifics of the forthcoming solution, at least by analogy with Landau levels, we consider an electron in a constant magnetic field as a particular case of a general electronic motion, which is described with an equation (see section 'Spin and Magnetic Moment' in Chapter 1)

$$\left(\gamma^\mu\left(p_\mu + \frac{e}{c}A_\mu\right) - mc\right)\psi = 0,$$

in which

$$p_\mu = (p_0, \ -\mathbf{p}) = \left(\frac{E}{c}, \mathrm{i}\hbar\frac{\partial}{\partial \mathbf{r}}\right)$$

is four-momentum,

$$A_\mu = (0, 0, \ -A_y(x), 0)$$

is four-potential of the electromagnetic field, E is the energy of the electron, e is the absolute charge of the electron and m is its mass. One supposes that the magnetic field vector is directed along z-axis and its magnitude has a weak dependence on x, almost constant. In a sense, the Dirac electron in a weakly inhomogeneous magnetic field is a curious anharmonicity effect.

To begin our calculations, we choose Dirac matrices; let these be

$$\gamma^0 = \begin{pmatrix} I & 0 \\ 0 & -I \end{pmatrix}, \quad \gamma^i = \begin{pmatrix} 0 & \sigma^i \\ -\sigma^i & 0 \end{pmatrix},$$

in which σ^i are 2×2 Pauli matrices,

$$\sigma^x = \begin{pmatrix} 0 & 1 \\ 1 & 0 \end{pmatrix}, \quad \sigma^y = \begin{pmatrix} 0 & -\mathrm{i} \\ \mathrm{i} & 0 \end{pmatrix}, \quad \sigma^z = \begin{pmatrix} 1 & 0 \\ 0 & -1 \end{pmatrix};$$

and I is a 2×2 unit matrix. Thus, our equation has a form

$$\left\{\gamma^0(E/c) - \gamma^x p_x - \gamma^y\left(p_y + \frac{e}{c}A_y\right) - \gamma^z p_z - mc\right\}\begin{pmatrix} \psi_{1,2} \\ \psi_{3,4} \end{pmatrix} = 0.$$

For brevity, we have here introduced a typical notation for a spinor as

$$\psi_{1,2} = \begin{pmatrix} \psi_1 \\ \psi_2 \end{pmatrix} \text{ and } \psi_{3,4} = \begin{pmatrix} \psi_3 \\ \psi_4 \end{pmatrix}.$$

Taking into account that

$$\gamma^0\psi = \begin{pmatrix} \psi_{1,2} \\ -\psi_{3,4} \end{pmatrix}, \quad \gamma^i\psi = \begin{pmatrix} \sigma^i\psi_{3,4} \\ -\sigma^i\psi_{1,2} \end{pmatrix},$$

and further that

$$\sigma^x\psi_{1,2} = \begin{pmatrix} \psi_2 \\ \psi_1 \end{pmatrix}, \quad \sigma^y\psi_{1,2} = \mathrm{i}\begin{pmatrix} -\psi_2 \\ \psi_1 \end{pmatrix}, \quad \sigma^z\psi_{1,2} = \begin{pmatrix} \psi_1 \\ -\psi_2 \end{pmatrix},$$

in an analogous manner,

$$\sigma^x \psi_{3,4} = \begin{pmatrix} \psi_4 \\ \psi_3 \end{pmatrix}, \quad \sigma^y \psi_{3,4} = \mathrm{i}\begin{pmatrix} -\psi_4 \\ \psi_3 \end{pmatrix}, \quad \sigma^z \psi_{3,4} = \begin{pmatrix} \psi_3 \\ -\psi_4 \end{pmatrix},$$

we obtain the system of equations for the spinor components:

$$\frac{E}{c}\psi_1 - p_x\psi_4 + \mathrm{i}\left(p_y + \frac{e}{c}A_y\right)\psi_4 - p_z\psi_3 - mc\psi_1 = 0,$$

$$\frac{E}{c}\psi_2 - p_x\psi_3 - \mathrm{i}\left(p_y + \frac{e}{c}A_y\right)\psi_3 + p_z\psi_4 - mc\psi_2 = 0,$$

$$-\frac{E}{c}\psi_3 + p_x\psi_2 - \mathrm{i}\left(p_y + \frac{e}{c}A_y\right)\psi_2 + p_z\psi_1 - mc\psi_3 = 0,$$

$$-\frac{E}{c}\psi_4 + p_x\psi_1 + \mathrm{i}\left(p_y + \frac{e}{c}A_y\right)\psi_1 - p_z\psi_2 - mc\psi_4 = 0.$$

Through a condition that potential A_y depends on x only, the solution becomes chosen in a form

$$\psi_j = u_j(x)\cdot \exp\left(\frac{\mathrm{i}p_y y}{\hbar} + \frac{\mathrm{i}p_z z}{\hbar}\right).$$

As a result, we have

$$f^- u_1 + \mathrm{i}\hbar\frac{\partial u_4}{\partial x} + \mathrm{i}Pu_4 - p_z u_3 = 0,$$

$$f^- u_2 + \mathrm{i}\hbar\frac{\partial u_3}{\partial x} - \mathrm{i}Pu_3 + p_z u_4 = 0,$$

$$-f^+ u_3 - \mathrm{i}\hbar\frac{\partial u_2}{\partial x} - \mathrm{i}Pu_2 + p_z u_1 = 0,$$

$$-f^+ u_4 - \mathrm{i}\hbar\frac{\partial u_1}{\partial x} + \mathrm{i}Pu_1 - p_z u_2 = 0,$$

in which $f^{\pm} = (E/c) \pm mc$, $P = p_y + (e/c)A_y$, p_y and p_z are c-numbers. We express u_3 and u_4 from the latter two equations,

$$u_3 = \frac{1}{f^+}\left(-\mathrm{i}\hbar\frac{\partial u_2}{\partial x} - \mathrm{i}Pu_2 + p_z u_1\right), \quad u_4 = \frac{1}{f^+}\left(-\mathrm{i}\hbar\frac{\partial u_1}{\partial x} + \mathrm{i}Pu_1 - p_z u_2\right),$$

and substitute them into the former two equations; as a result,

$$\frac{\partial^2 u_1}{\partial x^2} + \frac{1}{\hbar^2}\left(f^- f^+ - p_z^2 - P^2 - \hbar\frac{\partial P}{\partial x}\right)u_1 = 0,$$

$$\frac{\partial^2 u_2}{\partial x^2} + \frac{1}{\hbar^2}\left(f^- f^+ - p_z^2 - P^2 + \hbar\frac{\partial P}{\partial x}\right)u_2 = 0.$$

Equations for u_3 and u_4 yield the same result:

$$\frac{\partial^2 u_3}{\partial x^2} + \frac{1}{\hbar^2}\left(f^- f^+ - p_z^2 - P^2 - \hbar\frac{\partial P}{\partial x}\right)u_3 = 0,$$

$$\frac{\partial^2 u_4}{\partial x^2} + \frac{1}{\hbar^2}\left(f^- f^+ - p_z^2 - P^2 + \hbar\frac{\partial P}{\partial x}\right)u_4 = 0.$$

These equations correspond to negative values of energy and two possible projections of the electron spin.

We eventually combine equations to obtain

$$\frac{\partial^2 u}{\partial x^2} + \frac{1}{\hbar^2}\left(\frac{E^2}{c^2} - m^2c^2 - p_z^2 - \left(p_y + \frac{e}{c}A_y\right)^2 - \hbar e\frac{\sigma}{c}\frac{\partial A_y}{\partial x}\right)u = 0,$$

in which $\sigma = +1$ for components u_1 and u_3, $\sigma = -1$ for components u_2 and u_4.

We apply a condition that the magnitude of the field is almost constant and expand A_y in powers of x: $A_y = a_1x + a_2x^2 + \cdots = \sum_{i>0} a_i x^i$. Coefficients a_i for $i > 1$ are assumed to be sufficiently small that in zero-order approximation the z-projection of the magnetic-field vector equals $\partial A_y/\partial x = a_1 = \text{const}$. Moreover, we introduce by definition a function

$$F(x) = \left(p_y + \frac{e}{c}A_y\right)^2 + \hbar e\frac{\sigma}{c}\frac{\partial A_y}{\partial x},$$

and transform $F(x)$:

$$p_y + \frac{e}{c}A_y = \sum_i b_i x^i, \quad b_0 = p_y, \quad b_{i>0} = e\frac{a_i}{c};$$

$$\left(p_y + \frac{e}{c}A_y\right)^2 = \left(\sum_i b_i x^i\right)^2 = \sum_k B_k x^k, \quad B_k = b_0b_k + b_1b_{k-1} + b_2b_{k-2} + \cdots + b_kb_0;$$

$$\hbar e\frac{\sigma}{c}\frac{\partial A_y}{\partial x} = \hbar e\frac{\sigma}{c}\sum_k (k+1)a_{k+1}x^k.$$

Therefore, if $C_k = B_k + \hbar e(\sigma/c)(k+1)a_{k+1}$, then $F(x) = \sum_k C_k x^k$.

Let small coefficients a_i with $i > 1$ be coefficients of an anharmonic type such that there is a linear transformation $q = x - x_0$, then

$$F(x) = \sum_k C_k x^k = \sum_k Q_k q^k = F(q),$$

with $\partial F/\partial x|_{x=x_0} = 0$, i.e. $Q_1 = 0$. The arbitrary coefficients have a form

$$Q_k = \frac{1}{k!}F^{(k)}(q)_{q=0} = \frac{1}{k!}F^{(k)}(x)_{x=x_0} = \frac{1}{k!}\sum_j C_j j(j-1)\cdots(j-k+1)x_0^{j-k}.$$

We see that $Q_1 = \sum_j C_j j x_0^{j-1} = 0$, from which equation follows the value of x_0, then we find $Q_0 = \sum_j C_j x_0^j$ and other quantities Q_k.

We thus obtain equation

$$\frac{\partial^2 u}{\partial q^2} + \left(\kappa - \frac{Q_2}{\hbar^2}q^2 - \frac{Q_3}{\hbar^2}q^3 - \frac{Q_4}{\hbar^2}q^4 - \cdots\right)u = 0,$$

$$\kappa = \frac{1}{\hbar^2}\left(\frac{E^2}{c^2} - m^2c^2 - p_z^2 - Q_0\right).$$

One should primarily investigate a particular solution $a_{i>1} = 0$; in this case,

$$b_0 = p_y, \quad b_1 = e\frac{a_1}{c}, \quad B_0 = p_y^2, \quad B_1 = 2e\frac{a_1}{c}p_y, \quad B_2 = \left(\frac{ea_1}{c}\right)^2,$$

$$C_0 = B_0 + \hbar e\frac{\sigma}{c}a_1, \quad C_1 = B_1, \quad C_2 = B_2.$$

Furthermore,

$$Q_1 = C_1 + 2C_2x_0 = 0;$$

therefore,

$$x_0 = -\frac{cp_y}{ea_1},$$

and

$$Q_0 = C_0 + C_1x_0 + C_2x_0^2 = \hbar e\frac{\sigma}{c}a_1, \quad Q_2 = C_2 = \left(\frac{ea_1}{c}\right)^2;$$

other coefficients are equal to zero. Assuming $q=\sqrt{c\hbar/ea_1}\xi$, we obtain the Hermite equation

$$\frac{\partial^2 u}{\partial \xi^2}+\left(\frac{c\hbar}{ea_1}\kappa-\xi^2\right)u=0,$$

which gives that $(c\hbar/ea_1)\kappa=2n+1$; hence,

$$E^2=m^2c^4+c^2p_z^2+c\hbar ea_1(2n+1+\sigma),\quad n=0,1,2,\ldots,$$

in which $\sigma=+1$ for components u_1 and u_3, $\sigma=-1$ for components u_2 and u_4; a_1 is the strength of the magnetic field. All functions u_1, u_2, u_3 and u_4 become expressible through Hermite polynomials.

The exact solution of the Dirac equation for an electron in a homogeneous magnetic field is thus derived. The expression for energy includes two classical terms $(mc^2)^2$ and $(cp_z)^2$, plus a quantized quantity, which appears through the motion in the plane perpendicular to z-axis. The electron momentum directed along the magnetic-field vector retains continuous values, and the rotatory motion in plane xy is described with energy levels of harmonic-oscillator type.

To return to our general problem, we take into account the anharmonicity. Introducing new variable $q=(\hbar/\sqrt{Q_2})^{1/2}\xi\equiv\lambda\xi$, we have

$$-\frac{\partial^2 u}{\partial \xi^2}+\xi^2 u+\sum_{k>2}\frac{Q_k}{\hbar^2}\lambda^{k+2}\xi^k u=(\lambda^2\kappa)u.$$

Assuming

$$H^0=-\frac{\partial^2}{2\partial\xi^2}+\frac{1}{2}\xi^2,\quad c_p=\frac{Q_{p+2}}{2\hbar^2}\lambda^4=\frac{Q_{p+2}}{2Q_2},\quad \varepsilon=\frac{1}{2}(\lambda^2\kappa)=\frac{\hbar}{2\sqrt{Q_2}}\kappa,$$

obviously,

$$\left(H^0+\sum_{p>0}c_p\lambda^p\xi^{p+2}\right)u=\varepsilon u.$$

This equation is typical in the theory of anharmonicity; as $Q_2 \gg Q_3$, $Q_4,\ldots$, the solution might be expressible through a series of perturbation theory, for instance, in a form of polynomials of quantum numbers. The zero-order approximation is here a simple harmonic oscillator. In this case, $c_{p>0}=0$, $\varepsilon_0=n+1/2$, $n=0$, 1, 2,..., and functions $u(\xi)$ are expressed in terms of Hermite polynomials. From the general structure of an anharmonic Hamiltonian, arbitrary corrections to ε_0 clearly depend only on quantum number n in a polynomial manner. Each correction ε_α of order α ($\alpha=1$, 2,...) is hence a sum of some polynomials, which depend on coefficients c_p parametrically. These corrections add to unperturbed quantity ε_0, which represents the square of the energy; in the result we obtain some expansion in n or $n + 1/2$, i.e.

$$\varepsilon = \sum_i \zeta_i (n+1/2)^i.$$

Coefficients ζ_i are defined from a general solution with perturbation theory, which is given by a simple expression

$$\varepsilon = \varepsilon_0 + \sum_\alpha \lambda^\alpha \varepsilon_\alpha, \quad \varepsilon_\alpha = \frac{1}{\alpha} \sum_{(p\beta\gamma)\alpha} p c_p \Pi_{\beta\gamma}^{p+2}(n,n),$$

$$p = 1, 2, \ldots,$$

$$\beta, \gamma = 0, 1, 2, \ldots,$$

in which $(p\beta\gamma)\alpha$ denotes a summation over indices p, β, γ under a condition that $p + \beta + \gamma = \alpha$. Quantities $\Pi_{\beta\gamma}^s(n,n)$ are polynomials of quantum numbers as before, which follows from the recurrence relations and can be taken from the table. For instance, the first-order correction equals zero; the second-order correction is $\varepsilon_2 = -(30c_1^2 - 6c_2)(n+1/2)^2 + \text{const.}$, etc.

In conclusion, we note that one might consider the Klein–Fock–Gordon equation in a similar manner. For a particle with zero spin and charge e' in a weak inhomogeneous magnetic field, this equation has a form

$$\left(\frac{E^2}{c^2} - m^2c^2 - p_x^2 - \left(p_y - \frac{1}{c}e'A_y\right)^2 - p_z^2\right)\psi = 0.$$

Having put $p_x = -\mathrm{i}\hbar\partial/\partial x$, $p_y = \mathrm{p}_y$, $p_z = \mathrm{p}_z$, we obtain

$$\frac{\partial^2 \varphi}{\partial x^2} + \frac{1}{\hbar^2}\left(\frac{E^2}{c^2} - m^2c^2 - p_z^2 - P^2\right)\varphi = 0, \quad \psi = \varphi(x)\exp\left(\frac{\mathrm{i}p_y y}{\hbar} + \frac{\mathrm{i}p_z z}{\hbar}\right).$$

This equation is a particular case of the equation in Dirac theory considered above, here simply $\sigma = 0$. Other calculations remain valid.

The Resonance Interaction

Notable effects of the anharmonicity of molecular vibrations are displayed in infrared absorption spectra of liquids and crystals with molecular solutes. For instance, at low temperatures the dipole–dipole interaction becomes dominant for molecular impurities with a large first derivative of dipolar moment. This resonance interaction is responsible for the formation of bands in the infrared absorption spectra of low-temperature liquids SF_6, CF_4, NF_3, OCS and alkali-halide crystals with impurity defects of type XH–XH, in which X = O, S, Se, Te. The condition necessary for the resonance is here the degeneracy of the energy levels. Through the interaction, the initially coincident levels are shifted so that the degeneracy becomes eliminated, and the splittings of vibrational levels are observed in the spectra.

These problems, being stationary, generally begin from the equation on eigenvalues E and eigenfunctions $|\Psi\rangle$,

$$(H^0 + W)|\Psi\rangle = E|\Psi\rangle,$$

in which H^0 is a known Hamiltonian of a system under consideration and W is the interaction operator. Quantity W implies a perturbation, for instance, the dipole–dipole potential or simple anharmonic potential. Function $|\Psi\rangle$ is represented in a form of expansion,

$$|\Psi\rangle = \sum_j C_j|\Phi, j\rangle,$$

in which known functions $|\Phi, j\rangle$ correspond to degenerate levels E_j^0 of unperturbed Hamiltonian H^0. As a result,

$$\sum_j C_j(E_j^0 + W - E)|\Phi, j\rangle = 0,$$

from which

$$\sum_j C_j(W_{ij} - E\delta_{ij}) = 0, \quad i = 1, 2, \ldots, z; \tag{4.3}$$

W_{ij} denotes an element

$$E_j^0\delta_{ij} + \langle \Phi, i\,|W|\Phi, j\rangle,$$

and z is the number of degenerate levels. For system (4.3) to have a non-trivial solution, its determinant must equal zero, i.e.

$$\det(W_{ij} - E\delta_{ij}) = 0.$$

This equation, called secular, represents an algebraic equation of order z; its solution is sought for the energy levels that exhibit the influence of the interaction.

The correct wave functions are defined by system (4.3) and the condition that the integrated product of each wave function equals unity. After the solution of secular equation is obtained, system (4.3) becomes linearly dependent; one must therefore eliminate one equation. Eliminating the latter, we rewrite Eq. (4.3) in a form

$$\sum_{j=1}^{z-1} C_j(W_{ij} - E\delta_{ij}) = -\, C_z W_{iz}, \quad i = 1, 2, \ldots, z-1.$$

We multiply each such equation by $(W - E\delta)^{-1}{}_{ki}$, sum over index i, and obtain

$$C_k = (-1)^{z-k} \frac{\det_k(W - E\delta)}{\det_z(W - E\delta)} C_z,$$

in which $\det_k(W - E\delta)$ is given by the expression

$$\begin{vmatrix} W_{11} - E & \cdots & \begin{bmatrix} W_{1k} \\ W_{2k} \\ \vdots \\ W_{z-1,k} \end{bmatrix} & \cdots & (W-E\delta)_{1,z-1} & W_{1z} \\ W_{21} & \cdots & & \cdots & (W-E\delta)_{2,z-1} & W_{2z} \\ \vdots & \ddots & & \ddots & \vdots & \vdots \\ W_{z-1,1} & \cdots & & \cdots & (W-E\delta)_{z-1,z-1} & W_{z-1,z} \end{vmatrix}$$

as column k is deleted. Taking into account the condition of normalization, we define C_z that is accurate within a sign and then find other coefficients C_k,

$$C_k = \pm (-1)^{z-k} \det_k(W - E\delta) \left(\sum_{i=1}^{z} \det_i^2 (W - E\delta) \right)^{-1/2},$$

in which each value of E represents the solution of a secular equation.

Appealing to the history of the problem of vibrational resonances, we note that Fermi first recognized this phenomenon on studying the vibrational modes of linear symmetric molecule CO_2. This molecule has three normal modes of vibration, two of which are non-degenerate with frequencies ω_1 and ω_3, and one of which is doubly degenerate with frequency ω_2. As the frequency of the totally symmetric vibration ω_1 is almost equal to twice the frequency of degenerate vibration ω_2, a resonance might occur between levels ω_1 and $2\omega_2$ of the same symmetry, caused by anharmonicity of the vibrations. The values of energy levels shifted by anharmonicity are defined through a solution of this secular equation,

$$\begin{matrix} \langle n_1 + 1, n_2, n_3| \\ \langle n_1, n_2 + 2, n_3| \end{matrix} \begin{vmatrix} \hbar\omega_1 + W_{11} - E & W_{12} \\ W_{21} & 2\hbar\omega_2 + W_{22} - E \end{vmatrix} = 0,$$

in which $|n_1, n_2, n_3\rangle$ is the state vector that equals the product of three harmonic-oscillator functions with quantum numbers n_1, n_2 and n_3, which correspond to normal coordinates q_1, q_2 and q_3 of the molecule. The values of E are reckoned from quantity

$$\hbar\omega_1(n_1 + 1/2) + \hbar\omega_2(n_2 + 1) + \hbar\omega_3(n_3 + 1/2),$$

that represents the energy of the harmonic vibrations of CO_2. To understand the resonance phenomenon, we take into account in W only that part of the anharmonic potential that includes simultaneously variables q_1 and q_2. Restricting to the first order in the perturbation, W becomes expressible in a form $aq_1q_2^2$. In this case, diagonal matrix elements W_{11} and W_{22} equal zero, whereas

$$W_{12} = a\langle n_1 + 1, n_2, n_3 | q_1q_2^2 | n_1, n_2 + 2, n_3\rangle = (a/2\sqrt{2})\sqrt{(n_1 + 1)(n_2 + 1)(n_2 + 2)},$$

and $W_{12} = W_{21}$. Hence,

$$E_{1,2} = \frac{1}{2}(\hbar\omega_1 + 2\hbar\omega_2) \pm \frac{1}{2}\sqrt{(\hbar\omega_1 - 2\hbar\omega_2)^2 + (a^2/2)(n_1 + 1)(n_2 + 1)(n_2 + 2)}.$$

The sought splitting is defined by difference $(E_1 - E_2)$ and, at the condition of exact resonance $\hbar\omega_1 = 2\hbar\omega_2$, becomes

$$\delta_F = (a/\sqrt{2})\sqrt{(n_1 + 1)(n_2 + 1)(n_2 + 2)}.$$

One sees that the greater the extent of interaction between the levels, the greater the extent of splitting δ_F. It is convenient to generalize and write the Fermi splitting in a form $\delta_F = \sqrt{\zeta^2 + 4W_{12}^2}$, in which ζ is the spacing between levels with no interaction, which leads to elimination of the degeneracy. Considering highly excited vibrations, one should bear in mind the significant modification of the anharmonic part of the Hamiltonian; to obtain energy shifts correctly, one must take into account approximations of greater order with perturbation theory. In this case, one might increase the number of degenerate states; it thus becomes necessary to consider much more complicated secular equations, which correspond to Fermi resonance between excited vibrational states.

Dimers in Low-Temperature Liquids

Like Fermi resonance, the phenomenon of the splitting of vibrational levels is observed in the spectra of low-temperature liquids [45]. In this case, interacting vibrational levels belong to separate molecules, a fixed small distance apart. The interaction of these two molecules produces a dimer; we thus treat dimers in absorption, not single molecules – monomers. To elucidate this effect, we evaluate the splittings of the energy levels of dimers SF_6–SF_6 in liquid argon at temperature 93 K [62], considering only non-degenerate vibration $\omega_1(A_1)$, which maintains the symmetry of the molecule, and a triply degenerate vibration of type F_{1u} with frequency ω_3 [63].

Let the wave functions of the system of free molecules

$$|n_{1q}n_{1x}n_{1y}n_{1z}, n_{2q}n_{2x}n_{2y}n_{2z}\rangle$$

be products of harmonic vectors

$$|n_{1q}, n_{1x}, n_{1y}, n_{1z}\rangle \text{ and } |n_{2q}, n_{2x}, n_{2y}, n_{2z}\rangle.$$

Normal coordinate q and quantum number n_q correspond to vibrational mode ω_1; the triply degenerate oscillator with dimensionless coordinates x, y, z and quantum numbers n_x, n_y, n_z corresponds to vibrational mode ω_3. Subscripts 1 and 2 on quantum numbers denote the number of molecules in the dimer. Harmonic Hamiltonian H^0 describes the zero-order approximation; its eigenvalues are equal to a sum of vibrational energies of two molecules

$$E^0(n_{1q}n_{1x}n_{1y}n_{1z}) \text{ and } E^0(n_{2q}n_{2x}n_{2y}n_{2z}).$$

To take into account the interaction, we introduce Hamiltonian

$$H = H^0 + V(x_1, y_1, z_1; x_2, y_2, z_2) + W(q_1q_2, x_1x_2, y_1y_2, z_1z_2).$$

Here, V defines the perturbation caused by the induced interaction for individual levels of molecules in the dimer,

$$V = -a\frac{\alpha}{R^3}(x_1^2 + y_1^2 + 4z_1^2 + x_2^2 + y_2^2 + 4z_2^2).$$

In essence, this interaction being of van der Waals type corresponds to the interaction between a dipole of SF_6 and an induced dipole of Ar or SF_6. Perturbation W is the dipole–dipole interaction of molecules SF_6,

$$W = -2a(2z_1z_2 - x_1x_2 - y_1y_2) - 4bq_1q_2(2z_1z_2 - x_1x_2 - y_1y_2).$$

Here, α is the polarizability of SF_6, a and b are coefficients that have a linear dependence on R^{-3} and the squared first derivative of the dipolar moment of this molecule and R is the distance directed along z-axis between the molecules. Through this dipole–dipole interaction, one might infer that dimers manifest themselves in absorption spectra. The value of the first derivative d_3' (with respect to the normal coordinate, which number is indicated by subscript 3) of molecule SF_6 is 0.55 in debye units and estimated distance R is about 5 Å.

If we take only V into account, the degeneracy is partly removed, through a difference between expectation values of the parts of V containing x (or y) and z. To illustrate this effect, we calculate, with perturbation theory in the first order, the correction to the energy of each molecule for the triply degenerate vibration. Omitting the subscript, which would indicate the number of molecules, we have

$$\begin{aligned}\Delta E_V &= -a\frac{\alpha}{R^3}\langle n_q, n_x, n_y, n_z|(x^2+y^2+4z^2)|n_q, n_x, n_y, n_z\rangle \\ &= -a\frac{\alpha}{R^3}((n_x+1/2)+(n_y+1/2)+4(n_z+1/2)).\end{aligned}$$

Quantum numbers n_x (or n_y) and n_z appear in this expression for the correction in a non-equivalent manner; we thus have an altered energy, which is concerned with the appearance of the selected direction along z-axis,

$$\left(\hbar\omega_3 - 4a\frac{\alpha}{R^3}\right) - \left(\hbar\omega_3 - a\frac{\alpha}{R^3}\right) = -\,3a\frac{\alpha}{R^3}.$$

This result is obtainable in another manner. For instance, one should rewrite the anharmonic potential for each molecule taking V into account and then redefine quantity ω_3 for $x(y)$ and z correspondingly.

A further shift of levels occurs through dipole–dipole interaction W. In this case, the energy levels of the first-order approximation in a perturbation are defined by secular equations. Despite the influence of V, the matrix elements of quantity W are calculated between degenerate states. This approximation is appropriate because W and V possess the same order of smallness, so there is no necessity to involve the wave functions of first order in V to calculate the matrix elements. As an example, to find the energy levels we consider the splitting of level $\omega_1+\omega_3$. In this case, we treat four states,

$$\begin{aligned}&|11,00\rangle \quad (n_{1q}=1,\ n_{1z}=1,\ n_{2q}=0,\ n_{2z}=0),\\ &|00,11\rangle \quad (n_{1q}=0,\ n_{1z}=0,\ n_{2q}=1,\ n_{2z}=1),\\ &|10,01\rangle \quad (n_{1q}=1,\ n_{1z}=0,\ n_{2q}=0,\ n_{2z}=1),\\ &|01,10\rangle \quad (n_{1q}=0,\ n_{1z}=1,\ n_{2q}=1,\ n_{2z}=0),\end{aligned}$$

in which we omit quantum numbers that correspond to x- and y-components and equal zero in expressions for eigenstates. Calculating elementarily matrix elements

$$\langle 11,00|W|00,11\rangle = -8b\langle 1|q_1|0\rangle\langle 1|z_1|0\rangle\langle 0|q_2|1\rangle\langle 0|z_2|1\rangle = -2b,$$

$$\langle 11,00|W|10,01\rangle = -4a\langle 1|z_1|0\rangle\langle 0|z_2|1\rangle = -2a$$

and so on, and reckoning the energy from the level $\omega_1+\omega_3$, one might represent the secular equation for z-components in a form

$$\begin{array}{l} \langle 11,00| \\ \langle 00,11| \\ \langle 10,01| \\ \langle 01,10| \end{array}\begin{vmatrix} -\varepsilon & -2b & -2a & 0 \\ -2b & -\varepsilon & 0 & -2a \\ -2a & 0 & \zeta-\varepsilon & -2b \\ 0 & -2a & -2b & \zeta-\varepsilon \end{vmatrix} = 0,$$

in which ζ, having a magnitude of a few cm^{-1}, represents the difference between unperturbed energy values of states $|10,01\rangle$ and $|11,00\rangle$ with regard to only the mechanical anharmonicity. Having exactly solved this equation, we obtain values ε for the degenerate level shifts in a form

$$\varepsilon_{1,2,3,4} = (1/2)\left(\zeta \pm \sqrt{\zeta^2 + 16a^2}\right) \mp 2b.$$

If the order of a is a few cm^{-1}, b is much smaller and of order only 0.01 cm^{-1}. We have therefore exactly two non-coincident levels, which formally remain degenerate.

For the fundamental band ω_3, the influence of the dipole–dipole interaction yields the simplest secular equations

$$\begin{matrix}\langle 1,0|_x \\ \langle 0,1|_x\end{matrix}\begin{vmatrix} -\varepsilon & a \\ a & -\varepsilon \end{vmatrix} = 0 \text{ and } \begin{matrix}\langle 1,0|_y \\ \langle 0,1|_y\end{matrix}\begin{vmatrix} -\varepsilon & a \\ a & -\varepsilon \end{vmatrix} = 0$$

for excited x- and y-components, and

$$\begin{matrix}\langle 1,0|_z \\ \langle 0,1|_z\end{matrix}\begin{vmatrix} -\varepsilon & -2a \\ -2a & -\varepsilon \end{vmatrix} = 0$$

for excited z-components. Here, we everywhere omit quantum numbers that equal zero in denoting state vectors. As a result, for shift ε, which is reckoned from level ω_3, we obtain non-degenerate values $\pm 2a$ and doubly degenerate values $\pm a$. The transitions to only levels with symmetric wave functions

$$(1/\sqrt{2})|1,0\rangle_{x(y,z)} + (1/\sqrt{2})|0,1\rangle_{x(y,z)}$$

are optically permitted.

Diatomic Dimers

We consider a general mechanism for the resonance splitting of energy levels. Our purpose is to define the frequencies and intensities of vibrational transitions, taking as an example a pair of diatomic molecules that occupy adjacent anion vacancies in an ionic crystal [44]. Despite the simplicity, this system allows one to enunciate a conclusion about the effect of the resonance interaction involving information from absorption spectra of overtone transitions. The dipole–dipole interaction is the focus of our attention here. The repulsive forces of type charge–charge, an induced interaction and an interaction of the type charge–dipole fail to explain the splittings observed in spectra, as these interactions cause a shift of all vibrational lines retaining the spacing between the lines invariant; this condition is clear, for instance, from the problem for dimer $SF_6{-}SF_6$ in liquid argon, which is considered in the preceding section.

We write the Hamiltonian for a pair of interacting molecules in a form

$$H = H_1 + H_2 + W,$$

in which H_1 and H_2 are Hamiltonians of the monomers. Quantity W that equals fd_1d_2 represents the energy of the dipole–dipole interaction. Here, d_1 and d_2 are the dipolar moments of the molecules; factor

$$f = \frac{(\mathbf{e}_1\mathbf{e}_2)R^2 - 3(\mathbf{e}_1\mathbf{R})(\mathbf{e}_2\mathbf{R})}{R^5}$$

determines the mutual orientation of the molecules, in which unit vectors $\mathbf{e}_1$ and $\mathbf{e}_2$ characterize the directions of the dipole moments and vector $\mathbf{R}$ connects the centres of mass of the molecules in the dimer. Each operator, H_1 and H_2, is a Hamiltonian of anharmonic type

$$H^0 + \hbar\omega \sum_{p>0} a_p \xi^{p+2}$$

with well-known eigenfunctions $|n)$ and eigenvalues

$$E_n = \hbar\omega\left(n + \frac{1}{2}\right) + \sum_{\alpha} \frac{\hbar\omega}{\alpha} \sum_{(p\beta\gamma)\alpha} p a_p \Pi^{p+2}_{\beta\gamma}(n, n);$$

H^0 is the Hamiltonian of the harmonic oscillator with vibrational frequency ω, ξ is the vibrational variable and a_p are force parameters.

The frequencies of internal vibrations of diatomic molecules are more than 10 times the frequencies of lattice vibrations; the molecules in the dimer may thus be considered to be free anharmonic oscillators, and the influence of the crystal is taken into account when specifying the frequencies of harmonic vibrations and the anharmonicity parameters, which are measured from the corresponding isolated XH ions residing in the crystal field. Since diatomic XH defects cannot rotate freely in the crystal, the oscillators are additionally fixed in a rigorous manner. The energy of the dipole–dipole interaction, which is approximately 1/100 times the energy of molecular vibrations, is considered a perturbation in this model. Our model eventually becomes correct at liquid-helium temperatures.

The wave function of the zero-order approximation denoted $|n_1n_2\rangle$ is expressible in a form of product eigenvectors $|n_1)$ and $|n_2)$ of Hamiltonians H_1 and H_2. Eigenvalues of sum

$$H_1 + H_2$$

represent the unperturbed energy values of the dimer,

$$E_i^0 = \varepsilon_{n_1} + \varepsilon_{n_2};$$

each quantity, ε_{n_1} and ε_{n_2}, is given by an expression for the energy of an anharmonic oscillator:

$$\varepsilon_n = \hbar\omega(n + 1/2) - \hbar\omega x_e(n+1/2)^2 + \cdots.$$

Here, x_e is the anharmonicity parameter and index i denotes the number of degenerate (near each other) vibrational levels, which are characterized by quantum numbers n_1 and n_2; indices 1 and 2 correspond to molecules in the dimer. Coefficients a_p are expressible through Morse's potential with parameters D and a_M (see section 'Future Developments' in Chapter 3; $\lambda = 1$), i.e.

$$\varepsilon_n = \hbar\omega(n + 1/2) - \hbar\omega x_e(n+1/2)^2,$$

$$x_e = \frac{1}{2}a_M^2 = \frac{\hbar\omega}{4D}, \quad a_1 = -\frac{1}{4}\sqrt{x_e}, \quad a_2 = \frac{7}{48}x_e, \ldots.$$

This choice for the potential is highly appropriate because our initial data for each molecule are ω and x_e.

According to the results of the preceding section, energy levels E of the dimer with regard to the perturbation are defined in a secular equation:

$$\det(W_{ij} - E\delta_{ij}) = 0.$$

Matrix elements $\langle n_1n_2|W|m_1m_2\rangle$ of the dipole–dipole interaction correspond to quantities W_{ij}, in which quantum numbers n_1 and n_2 characterize the state with energy E_i^0, with m_1 and m_2 for the state with E_j^0. As vibrational wave functions have real values, the interaction matrix is symmetric. Obviously,

$$\langle n_1n_2|W|m_1m_2\rangle = f(n_1|d_1|m_1)(n_2|d_2|m_2).$$

Using Eq. (3.7), for each matrix element, $(n_1|d_1|m_1)$ and $(n_2|d_2|m_2)$, we have

$$(n|d|m) = \sqrt{g_{nm}}\sum_{s\alpha} D_s\Pi^s_{(\beta\gamma)\alpha}(m-n), \quad D_s = \frac{1}{s!}2^{-s/2}d^{(s)},$$

in which derivatives $d^{(s)}$ are the corresponding coefficients in the expansion for the dipolar-moment function in terms of vibrational coordinate q; recall that $\xi = \sqrt{2}q$. Restricting our consideration to approximations $s \leq 1$ and $\alpha \leq 2$, choosing the proper polynomials from the table and making elementary calculations, we obtain non-zero matrix elements:

$$(n|d|n) = d^0 - 6\sqrt{2}a_1d'(n + 1/2),$$

$$(n|d|n+1) = d'(1 + (22a_1^2 - 6a_2)(n+1))\sqrt{g_{n,n+1}/2},$$

$$(n|d|n+2) = \sqrt{2}a_1d'\sqrt{g_{n,n+2}},$$

$$(n|d|n+3) = d'(3a_1^2 + a_2)\sqrt{g_{n,n+3}/2}.$$

As a result, the energy levels of the dimer represent the eigenvalues of the matrix of the dipole–dipole interaction, with elements that are determined by quantities ω, x_e, d^0 and d'.

We first consider the fundamental transition, i.e. the vibrational transition from the ground non-degenerate state $|00)$ with energy

$$E(0,0) = 2\varepsilon_0 + \langle 00|W|00\rangle$$

to the first excited level ($n_1 + n_2 = 1$); $|n_1 n_2)$ denotes the correct wave function of the dimer with regard to W. Note that quantity $E(0,0)$ is calculated in the first order according to perturbation theory for non-degenerate states. The first vibrational level, which is unperturbed by the dipole–dipole interaction, is doubly degenerate; states $|10\rangle$ and $|01\rangle$ correspond to this level. The energy shifts of the dimer are determined according to the simplest secular equation:

$$\begin{vmatrix} E_1^0 + \langle 10|W|10\rangle - E & \langle 10|W|01\rangle \\ \langle 10|W|01\rangle & E_2^0 + \langle 01|W|01\rangle - E \end{vmatrix} = 0.$$

The removal of this degeneracy hence leads to two values of energy,

$$E_{1,2} = \frac{1}{2}(W_{11} + W_{22} \pm \delta),$$

$$\delta = \sqrt{\zeta^2 + 4W_{12}^2}, \quad \zeta = W_{22} - W_{11},$$

$$W_{11} = E_1^0 + \langle 10|W|10\rangle, \quad W_{12} = \langle 10|W|01\rangle, \quad W_{22} = E_2^0 + \langle 01|W|01\rangle,$$

in which δ is the spacing between the interacting levels with regard to the dipole–dipole interaction; E_1^0 and E_2^0 are unperturbed levels of the dimer. Quantity $\delta/2$ determines the shifts of the levels of the fundamental transition upon its splitting, with these shifts being invariably symmetric with respect to level $(W_{11} + W_{22})/2$.

For each found value, E_1 and E_2, one must define a wave function,

$$|10) = C_{11}|10\rangle + C_{12}|01\rangle \text{ and } |01) = C_{21}|10\rangle + C_{22}|01\rangle,$$

in which expansion coefficients C_{i1} and C_{i2} satisfy system (4.3),

$$(W_{11} - E_i)C_{i1} + W_{12}C_{i2} = 0,$$

$$W_{12}C_{i1} + (W_{22} - E_i)C_{i2} = 0,$$

under the condition $C_{i1}^2 + C_{i2}^2 = 1$; $i = 1, 2$. Thus,

$$C_{11} = -C_{22} = \Sigma\sqrt{\frac{\delta - \zeta}{2\delta}}, \quad C_{12} = C_{21} = \sqrt{\frac{\delta + \zeta}{2\delta}}, \quad \Sigma = \frac{W_{12}}{|W_{12}|},$$

in the case of resonance, $W_{11} = W_{22}$; hence, $\zeta = 0$ and the obtained formulae become simple. As a result,

$$E_1 = W_{11} + |W_{12}|, \quad |10) = \frac{\Sigma|10\rangle + |01\rangle}{\sqrt{2}};$$

$$E_2 = W_{11} - |W_{12}|, \quad |01) = \frac{|10\rangle - \Sigma|01\rangle}{\sqrt{2}}.$$

There remains to be considered the question of the intensities of transitions. What probabilities have transitions of the dimer from the ground state to an excited state, for instance, $|10)$ or $|01)$? According to the general definition, in the dipole approximation, absorption intensity I is proportional to a product of the squared matrix element of the total dipolar moment,

$$\mathbf{e}_1 d_1 + \mathbf{e}_2 d_2,$$

and transition frequency ω_{0z},

$$I_{0z} \sim \omega_{0z}((00|d_1|n_1 n_2)^2 + (00|d_2|n_1 n_2)^2 + 2(00|d_1|n_1 n_2)(00|d_2|n_1 n_2)\cos\vartheta),$$

in which $|00)$ and $|n_1 n_2)$ are vectors of initial and final states, respectively; ϑ is the angle between $\mathbf{e}_1$ and $\mathbf{e}_2$ and $n_1 + n_2 = z$. Vector $|n_1 n_2)$ represents in general a linear combination,

$$|n_1 n_2) = \sum_{(m_1 m_2)z} C_{m_1 m_2} |m_1 m_2\rangle,$$

whereas $|00) \approx |00\rangle$. For each solution of the secular equation, coefficients $C_{m_1 m_2}$ are determined through system (4.3) and the condition that wave function $|n_1 n_2)$ has been normalized to unity. We calculate the matrix elements

$$(00|d_1|n_1 n_2) = \sum_{(m_1 m_2)z} C_{m_1 m_2}(0|d_1|m_1)(0|m_2) = C_{z0}(0|d_1|z),$$

and

$$(00|d_2|n_1 n_2) = C_{0z}(0|d_2|z).$$

Because of the identity of the molecules in the dimer, the matrix elements of functions d_1 and d_2 are equal to each other. Replacing d_1 and d_2 with d, we obtain

$$I_{0z} \sim \omega_{0z}(0|d|z)^2(C_{z0}^2 + C_{0z}^2 + 2C_{z0}C_{0z}\cos\vartheta). \tag{4.4}$$

This general expression determines the intensity of the vibrational transition of the dimer.

For example, considering the fundamental transition, we have

$$C_{z0} = \frac{\Sigma}{\sqrt{2}} \text{ and } C_{0z} = \frac{1}{\sqrt{2}} \text{ for } \omega_{01} = \frac{E_1 - E(0,0)}{\hbar},$$

$$C_{z0} = \frac{1}{\sqrt{2}} \text{ and } C_{0z} = -\frac{\Sigma}{\sqrt{2}} \text{ for } \omega_{01} = \frac{E_2 - E(0,0)}{\hbar}.$$

In this case, formula (4.4) becomes highly elegant,

$$I_{01} \sim \omega_{01}(0|d|1)^2(1 \pm \Sigma \cos \vartheta), \tag{4.5}$$

in which the plus sign corresponds to value E_1 and the minus to E_2. Thus, if the molecules in the dimer have parallel or antiparallel orientations, a transition is permitted to only one level; the transition to the other level is strictly forbidden.

On the Theory of Overtones

A dipole–dipole interaction between equivalent diatomic molecules yields a splitting of the energy levels of the dimer; this interaction thus displays the resonance character. The vibrational levels of the dimer are determined according to a secular equation. The first energy level is doubly degenerate; as a result, solving the secular equation, we obtain the first splitting. The next vibrational level is triply degenerate, and perturbed energy values are defined according to a third-order equation. As a rule, for the first overtone without interaction, we have three levels, two of which exactly coincide and the other is shifted from this level pair through mechanical anharmonicity. Through the dipole–dipole interaction, the degenerate levels become split and the non-degenerate initial level shifts by a value of the diagonal matrix element. The fourth-order equation corresponds to the second overtone transition and so on. This scheme is generally applicable to find arbitrary energy levels of the dimer.

One might not obtain the exact solution of secular equation in all possible cases at once, but one might always simplify the initial problem, for instance, with the aid of a convenient orthogonal transformation A, which preserves the eigenvalues of interaction matrix W. Having examined the solution for the fundamental transition, we readily guess that one must introduce this matrix,

$$A = \begin{pmatrix} \boldsymbol{\sigma} & \mathbf{0} \\ \mathbf{0} & \mathbf{1} \end{pmatrix}, \quad \sigma = \frac{1}{\sqrt{2}}\begin{pmatrix} 1 & 1 \\ 1 & -1 \end{pmatrix},$$

in which **1** and **0** are unit and zero matrices, correspondingly. As a concrete case, we consider the degeneracy of the first two states. The secular equation for transformed matrix AWA', in which A' is the transposed matrix with regard to the symmetry of W_{ij}, is expressible in a form

$$\begin{vmatrix} \frac{1}{2}(W_{11}+2W_{12}+W_{22})-E & \frac{1}{2}(W_{11}-W_{22}) & \cdots & \frac{1}{\sqrt{2}}(W_{1z}+W_{2z}) \\ \frac{1}{2}(W_{11}-W_{22}) & \frac{1}{2}(W_{11}-2W_{12}+W_{22})-E & \cdots & \frac{1}{\sqrt{2}}(W_{1z}-W_{2z}) \\ \vdots & \vdots & \ddots & \vdots \\ \frac{1}{\sqrt{2}}(W_{1z}+W_{2z}) & \frac{1}{\sqrt{2}}(W_{1z}-W_{2z}) & \cdots & W_{zz}-E \end{vmatrix} = 0.$$

In the case of a resonance, one should specify

$$W_{11} = W_{22}.$$

From the above equation, the non-diagonal elements, which correspond to the degenerate states, therefore vanish.

Proceeding to a calculation of the overtone levels, in the case of the first overtone one must find the solution of secular equation

$$\begin{matrix} \langle 20| \\ \langle 11| \\ \langle 02| \end{matrix} \begin{vmatrix} W_{11}-E & W_{12} & W_{13} \\ W_{12} & W_{22}-E & W_{23} \\ W_{13} & W_{23} & W_{33}-E \end{vmatrix} = 0.$$

As the interacting molecules in the dimer are isotopically identical, their corresponding parameters ω, x_e, d^0 and d' are equal to each other. Hence, $W_{11} = W_{33}$, $W_{12} = W_{23}$, and one might apply transformation A for the degenerate states $|20\rangle$ and $|02\rangle$. As a result,

$$E_1 = W_{11} - W_{13},$$
$$E_{2,3} = E_0 + \frac{1}{2}(W_{13} \pm \delta),$$

in which

$$E_0 = \frac{1}{2}(W_{11} + W_{22}), \quad \delta = \sqrt{(\zeta - W_{13})^2 + 8W_{12}^2}, \quad \zeta = W_{22} - W_{11}.$$

Thus, on invoking the interaction, unperturbed E_1^0 and E_3^0 levels split, whereas non-degenerate E_2^0 level shifts through the perturbation by roughly a value $\langle 11|W|11\rangle$.

The effect of a resonance interaction on the third vibrational level is considered in an analogous manner. The solution of secular equation

$$\begin{matrix} \langle 30| \\ \langle 21| \\ \langle 12| \\ \langle 03| \end{matrix} \begin{vmatrix} W_{11}-E & W_{12} & W_{13} & W_{14} \\ W_{12} & W_{22}-E & W_{23} & W_{24} \\ W_{13} & W_{23} & W_{33}-E & W_{34} \\ W_{14} & W_{24} & W_{34} & W_{44}-E \end{vmatrix} = 0$$

with condition that $W_{11}=W_{44}$, $W_{22}=W_{33}$, $W_{12}=W_{34}$ and $W_{13}=W_{24}$ has the form

$$E_{1,2}=E_0+\frac{1}{2}(W_{14}+W_{23}\pm\delta_1),$$

$$E_{3,4}=E_0-\frac{1}{2}(W_{14}+W_{23}\pm\delta_2),$$

in which $E_0=(1/2)(W_{11}+W_{22})$, $\zeta=W_{22}-W_{11}$ and

$$\delta_1=\sqrt{(\zeta-W_{14}+W_{23})^2+4(W_{12}+W_{13})^2},$$

$$\delta_2=\sqrt{(\zeta+W_{14}-W_{23})^2+4(W_{12}-W_{13})^2}.$$

According to the above expressions, the splittings of perturbed levels E_1 and E_4, also E_2 and E_3, with respect to level E_0 shifted due to the dipole interaction are non-symmetric. Through this circumstance, the splittings of the overtone levels differ significantly from those of the fundamental transitions, for which an exact symmetry is observed.

We proceed to discuss higher overtones. The number of degeneracies generally coincides with the number of orthogonal transformations used to simplify the initial equation. The condition of the degeneracy of levels i and j is expressible in the form of an equality of the corresponding diagonal matrix elements, $W_{ii}=W_{jj}$. The conditions of the resonance are defined by equalities $\langle lm|W|ks\rangle=\langle sk|W|ml\rangle$. The secular equation is factorized by performing all orthogonal transformations, sorting the initially non-degenerate levels into individual groups. The analytic exact solutions thus exist minimally in the case of the first two overtones. For higher overtones, the factorization leads to equations of lower order, the approximate solutions of which are readily obtainable, for instance, by successive diagonalizations.

What is the impact of this approach? The transformation for the first overtone is $A_{13}WA'_{13}$, for the second overtone $A_{14}A_{23}WA'_{23}A'_{14}$, in which the subscripts on matrix A indicate the numbers of the degenerate states. To generalize, we represent the transformation for an arbitrary even overtone in a form

$$A_{1z}A_{2,z-1}\cdots A_{z/2,z/2+1}WA'_{z/2,z/2+1}\cdots A'_{2,z-1}A'_{1z},$$

accordingly, for an odd overtone

$$A_{1z}A_{2,z-1}\cdots A_{(z-1)/2,(z+3)/2}WA'_{(z-1)/2,(z+3)/2}\cdots A'_{2,z-1}A'_{1z},$$

in which z is the total number of energy levels, $z > 1$. We introduce matrices

$$K = A_{1z}A_{2,z-1}\cdots A_{z/2,z/2+1} \text{ and } L = A_{1z}A_{2,z-1}\cdots A_{(z-1)/2,(z+3)/2};$$

having calculated the products, one might represent them in an explicit form

$$K = \frac{1}{\sqrt{2}}\begin{pmatrix} \mathbf{1} & \boldsymbol{\kappa} \\ \boldsymbol{\kappa} & -\mathbf{1} \end{pmatrix}, \quad L = \frac{1}{\sqrt{2}}\begin{pmatrix} \mathbf{1} & \mathbf{0} & \boldsymbol{\kappa} \\ \mathbf{0} & \sqrt{2} & \mathbf{0} \\ \boldsymbol{\kappa} & \mathbf{0} & -\mathbf{1} \end{pmatrix}, \quad \boldsymbol{\kappa} = \begin{pmatrix} 0 & \cdots & 1 \\ \vdots & \ddots & \vdots \\ 1 & \cdots & 0 \end{pmatrix},$$

in which $\boldsymbol{\kappa}$ is a matrix with unity on the secondary diagonal, other elements equalling zero. Quantities K and L satisfy the next simple relations

$$K^2 = KK' = KK^{-1} = \frac{1}{2}(\mathbf{1} + \boldsymbol{\kappa}^2) = \mathbf{1}$$

and

$$L^2 = LL' = LL^{-1} = \frac{1}{2}\begin{pmatrix} \mathbf{1} + \boldsymbol{\kappa}^2 & \mathbf{0} & \mathbf{0} \\ \mathbf{0} & 2 & \mathbf{0} \\ \mathbf{0} & \mathbf{0} & \mathbf{1} + \boldsymbol{\kappa}^2 \end{pmatrix} = \mathbf{1}.$$

To consider the K-transformation in detail, we rewrite the interaction matrix for the case of an even overtone in a form

$$W = \begin{pmatrix} \mathbf{w}_{11} & \mathbf{w}_{12} \\ \mathbf{w}_{21} & \mathbf{w}_{22} \end{pmatrix}.$$

As matrices $\mathbf{w}_{ij}$ and $\boldsymbol{\kappa}$ have the same size, their size is not shown everywhere. On performing the transformation, we obtain

$$KWK' = \frac{1}{2}\begin{pmatrix} \mathbf{w}_{11} + \boldsymbol{\kappa}\mathbf{w}_{22}\boldsymbol{\kappa} + \mathbf{w}_{12}\boldsymbol{\kappa} + \boldsymbol{\kappa}\mathbf{w}_{21} & \mathbf{w}_{11}\boldsymbol{\kappa} - \boldsymbol{\kappa}\mathbf{w}_{22} + \boldsymbol{\kappa}\mathbf{w}_{21}\boldsymbol{\kappa} - \mathbf{w}_{12} \\ \boldsymbol{\kappa}\mathbf{w}_{11} - \mathbf{w}_{22}\boldsymbol{\kappa} + \boldsymbol{\kappa}\mathbf{w}_{12}\boldsymbol{\kappa} - \mathbf{w}_{21} & \boldsymbol{\kappa}\mathbf{w}_{11}\boldsymbol{\kappa} + \mathbf{w}_{22} - \boldsymbol{\kappa}\mathbf{w}_{12} - \mathbf{w}_{21}\boldsymbol{\kappa} \end{pmatrix}.$$

The condition of degeneracy levels,

$$\mathbf{w}_{11} = \boldsymbol{\kappa}\mathbf{w}_{22}\boldsymbol{\kappa},$$

and this equality,

$$\mathbf{w}_{12} = \boldsymbol{\kappa}\mathbf{w}_{21}\boldsymbol{\kappa},$$

which defines the resonance condition, join the latter expression. As a result,

$$KWK' = \begin{pmatrix} \mathbf{w}_{11} + \mathbf{w}_{12}\boldsymbol{\kappa} & \mathbf{0} \\ \mathbf{0} & \mathbf{w}_{22} - \mathbf{w}_{21}\boldsymbol{\kappa} \end{pmatrix},$$

and the secular equation for the transformed interaction matrix W is factorized into two equations, which have half the order with respect to the order of the initial equation. For example, in the case of the second overtone, the factorization leads to two second-order equations and we have an exact algebraic solution. The secular equation for states $|50\rangle$, $|41\rangle$, $|32\rangle$, $|23\rangle$, $|14\rangle$ and $|05\rangle$, which describes a fourth overtone, is resolved into two third-order equations

$$\begin{vmatrix} W_{33} + W_{34} - E & W_{23} + W_{24} & W_{13} + W_{14} \\ W_{23} + W_{24} & W_{22} + W_{25} - E & W_{12} + W_{15} \\ W_{13} + W_{14} & W_{12} + W_{15} & W_{11} + W_{16} - E \end{vmatrix} = 0$$

and

$$\begin{vmatrix} W_{33} - W_{34} - E & W_{23} - W_{24} & W_{13} - W_{14} \\ W_{23} - W_{24} & W_{22} - W_{25} - E & W_{12} - W_{15} \\ W_{13} - W_{14} & W_{12} - W_{15} & W_{11} - W_{16} - E \end{vmatrix} = 0$$

and so on.

In the case of an odd overtone, it is convenient to represent interaction matrix W in a form

$$W = \begin{pmatrix} \mathbf{w}_{11} & \mathbf{w}_{12} & \mathbf{w}_{13} \\ \mathbf{w}_{21} & W_{22} & \mathbf{w}_{23} \\ \mathbf{w}_{31} & \mathbf{w}_{32} & \mathbf{w}_{33} \end{pmatrix},$$

in which $\mathbf{w}_{12}$ and $\mathbf{w}_{32}$ are column matrices, $\mathbf{w}_{21}$ and $\mathbf{w}_{23}$ are row matrices. On making the L-transformation with conditions

$$\mathbf{w}_{11} = \boldsymbol{\kappa}\mathbf{w}_{33}\boldsymbol{\kappa} \text{ and } \mathbf{w}_{13} = \boldsymbol{\kappa}\mathbf{w}_{31}\boldsymbol{\kappa},$$

we obtain

$$LWL' = \begin{pmatrix} \mathbf{w}_{11} + \mathbf{w}_{13}\boldsymbol{\kappa} & (1/\sqrt{2})(\mathbf{w}_{12} + \boldsymbol{\kappa}\mathbf{w}_{32}) & \mathbf{0} \\ (1/\sqrt{2})(\mathbf{w}_{21} + \mathbf{w}_{23}\boldsymbol{\kappa}) & W_{22} & (1/\sqrt{2})(\mathbf{w}_{21}\boldsymbol{\kappa} - \mathbf{w}_{23}) \\ \mathbf{0} & (1/\sqrt{2})(\boldsymbol{\kappa}\mathbf{w}_{12} - \mathbf{w}_{32}) & \mathbf{w}_{33} - \mathbf{w}_{31}\boldsymbol{\kappa} \end{pmatrix}.$$

For overtones beyond the first, the factorization has a meaning only with definite additional conditions. From a physical point of view, one might simply exclude the non-degenerate level from consideration. For this purpose, we must neglect all elements of matrices $\mathbf{w}_{12}$, $\mathbf{w}_{21}$, $\mathbf{w}_{23}$ and $\mathbf{w}_{32}$, then

$$LWL' = \begin{pmatrix} \mathbf{w}_{11} + \mathbf{w}_{13}\boldsymbol{\kappa} & \mathbf{0} & \mathbf{0} \\ \mathbf{0} & W_{22} & \mathbf{0} \\ \mathbf{0} & \mathbf{0} & \mathbf{w}_{33} - \mathbf{w}_{31}\boldsymbol{\kappa} \end{pmatrix}$$

and the secular equation is exactly factorized. In turn, the non-degenerate level shifts by a quantity of the diagonal matrix element of interaction. For instance, considering the secular equation for states $|40\rangle$, $|31\rangle$, $|22\rangle$, $|13\rangle$ and $|04\rangle$, we obtain two exact energy values

$$E_{1,2} = E_0 - \frac{1}{2}(W_{15} + W_{24} \pm \delta_1),$$

in which

$$E_0 = \frac{1}{2}(W_{11} + W_{22}), \quad \delta_1 = \sqrt{(\zeta + W_{15} - W_{24})^2 + 4(W_{12} - W_{14})^2}, \quad \zeta = W_{22} - W_{11},$$

and the third-order equation

$$\begin{vmatrix} W_{22} + W_{24} - E & \sqrt{2}W_{23} & W_{12} + W_{14} \\ \sqrt{2}W_{23} & W_{33} - E & \sqrt{2}W_{13} \\ W_{12} + W_{14} & \sqrt{2}W_{13} & W_{11} + W_{15} - E \end{vmatrix} = 0,$$

which can be solved by assuming $E_3 = W_{33}$. Then

$$E_{4,5} = E_0 + \frac{1}{2}(W_{15} + W_{24} \pm \delta_2),$$

in which $\delta_2 = \sqrt{(\zeta - W_{15} + W_{24})^2 + 4(W_{12} + W_{14})^2}$.

XH–XH Pairs

For pairs of ions, the intensity of a spectral absorption is characterized by a quadratic dependence on the concentration of impurity molecules. These lines are shifted from lines of isolated defects about $10\ \mathrm{cm}^{-1}$ to smaller wavenumbers. Part of the shift of the lines, belonging to XH–XH, is due to the resonance dipole–dipole interaction. Another part corresponds to dimers XH–XD with no resonance and shifts arise through the static interaction. For instance, the typical wavenumbers of the internal vibration of SH differ about $1000\ \mathrm{cm}^{-1}$ from those of SD, and only a static interaction between these molecules leads to the shift of non-resonant lines observed in the spectra. The energy values of the non-resonant levels are readily obtainable with perturbation theory for non-degenerate states.

The effect of the dipole–dipole interaction for a pair of ions XH is taken into account in two stages: we consider initially the shift of degenerate levels, arising mainly due to the static dipole moment, and we then consider the interaction between the levels, causing their splitting. If the interaction between the levels is disregarded, all non-diagonal elements of the interaction matrix are set to zero; the energy levels then shift by the value of the diagonal element of the perturbation. In this case, energy

$$E(n_1, n_2) = \varepsilon_{n_1} + \varepsilon_{n_2} + \langle n_1 n_2 | W | n_1 n_2 \rangle$$

describes the static shift without removal of the degeneracy. As unperturbed levels,

$$E_i^0 = \varepsilon_{n_1} + \varepsilon_{n_2}$$

are included in matrix elements W_{ii}, $E(n_1, n_2)$ exactly coincides with W_{ii}. Taking into account the non-diagonal elements of the interaction matrix, we are led to the secular equations. Finding the solutions of these equations, we obtain the vibrational levels of the dimer; the degeneracies are eliminated. Subtracting energy $E(0,0)$ of the ground state of the dimer from these energy-level values, we obtain the sought transition wavenumbers of the diatomic dimers. In this case, it is convenient to reckon the splittings from degenerate level $E(n_1, n_2)$.

As an example, we consider the splitting of lines of diatomic dimer SH–SH in crystalline KCl. In a crystalline lattice, impurity defects SH substitute halide atoms, occupy the nearest anion vacancies and are oriented along a direction ⟨111⟩ either parallel or antiparallel to each other [43]. As the two negatively charged molecules approach each other, the pair appears at a distance 4.45 Å. Each ion is described by four parameters found from the spectra of isolated defects SH: frequency of harmonic vibrations ν_e (cm^{-1}), anharmonicity parameter x_e, and quantities d^0 and d' (in debye units). According to the literature [19,43,44], for ^{32}SH,

$$\nu_e = 2691.7, \quad x_e = 0.0187, \quad d^0 = 0.3, \quad d' = 0.214;$$

correspondingly for ^{34}SH,

$$\nu_e = 2689.3, \quad x_e = 0.0186, \quad d^0 = 0.3, \quad d' = 0.214.$$

The calculated transition wavenumbers of SH–SH dimers in crystalline KCl for the case of a parallel orientation of the molecules are presented in Table 4.1; the positions of the degenerate levels of ^{32}SH–^{32}SH are reckoned from level $E(0,0) =$ 2650.7 cm^{-1} [44]. The dimers formed by molecules with distinct isotopic composition ^{32}SH–^{34}SH are characterized by a quasi-exact resonance; in this case there is no exact degeneracy of unperturbed levels. Nevertheless, the corresponding shifts

Table 4.1 Transition Wavenumbers/cm^{-1} of SH–SH Dimer in KCl for the Case of a Parallel Orientation of the Molecules

Transitions	Degenerate Levels of ^{32}SH–^{32}SH	Vibrational Transition Wavenumbers of the Dimers	
		^{32}SH–^{32}SH	^{32}SH–^{34}SH
Fundamental	2580.6	2579.2	2578.0
	2580.6	2581.9	2581.2
First overtone	5060.8	5060.7	5057.4
	5060.8	5060.8	5060.8
	5160.9	5161.0	5159.1
Second overtone	7440.8	7440.7	7436.4
	7440.8	7440.7	7440.7
	7640.9	7638.2	7635.5
	7640.9	7643.6	7641.1

of the vibrational levels are readily obtainable through an analogous procedure of solving the secular equations [19,44]. For instance, one might apply the method of successive diagonalization of the interaction matrix or our formulae obtained for the case of the exact resonance.

We discern significant changes for the fundamental and second-overtone transitions of ^{32}SH–^{32}SH. Denoting the quantity of splitting of lines of the dimer as $\Delta_{0\rightarrow z}$, we have

$$\Delta_{0\rightarrow 1} = 2.7\ \text{cm}^{-1} \quad \text{for lines } |00)-|01) \text{ and } |00)-|10),$$

$$\Delta_{0\rightarrow 2} = 0.1\ \text{cm}^{-1} \quad \text{for lines } |00)-|02) \text{ and } |00)-|20),$$

$$\Delta_{0\rightarrow 3} = 5.4\ \text{cm}^{-1} \quad \text{for lines } |00)-|12) \text{ and } |00)-|21).$$

For transitions $|00)-|03)$ and $|00)-|30)$, the splitting is only 0.0023 cm^{-1}. The results obtained agree satisfactorily with the results of preliminary measurements performed at temperature 15 K and with an instrumental resolution 0.3 cm^{-1}; whereas the fundamental transitions in the pair ^{32}SH–^{32}SH reveal a splitting 2.7 cm^{-1}, the splitting for the first overtone is smaller than 0.3 cm^{-1}.

The selection rules for the permitted vibrational transitions are determined according to the calculated intensity values. According to Eq. (4.5),

$$I_{01} \sim \omega_{01}(0|d|1)^2(1 \pm \Sigma \cos \vartheta).$$

This formula specifies the selection rules for the fundamental transitions. In the case of a parallel orientation of molecules, $\vartheta = 0$ and $\Sigma = -1$. A transition is hence allowed to only one level; the transition to the other level is precisely forbidden. The allowed transition occurs to the state with the lower energy. All these specific features are experimentally confirmed [43].

For overtone lines, similar patterns of intensities I_{0z} are obtained. According to Eq. (4.4), the most intense transitions are $|00)-|02)$ and $|00)-|03)$. If quantities I_{01}, I_{02} and I_{03} denote the intensities of transitions $|00)-|01)$, $|00)-|02)$ and $|00)-|03)$, for $^{32}SH-^{32}SH$ dimer, we have

$$I_{01}/I_{02} = 50 \text{ and } I_{01}/I_{03} = 2000.$$

The combination transitions $|00)-|11)$, $|00)-|12)$ and $|00)-|21)$ turn out to be weak.

Concerning the characteristics of isolated defects, our theory thus enables one to evaluate the wavenumbers and intensities for vibrational transitions of diatomic dimers. For the case of a parallel orientation SH−SH in KCl, experiment convincingly conforms to the result of a direct calculation. Vibrational levels of SH−SH, for the case $\vartheta = 0$, lie lower than the levels corresponding to a non-parallel orientation of molecules; calculated wavenumbers $\nu_{0\to z}$ for the fundamental $(0 \to 1)$ and first overtone $(0 \to 2)$ transitions $^{32}SH-^{32}SH$ (see Table 4.1) agree satisfactorily with values

$$\nu_{0\to 1} = 2579.21 \text{ cm}^{-1} \text{ and } \nu_{0\to 2} = 5059.81 \text{ cm}^{-1},$$

found from experiment [19,43]. Both these conclusions confirm that we have correctly chosen configuration $\vartheta = 0$ for the defects in the pair. One might expect that the molecules forming the dimer have slightly varied parameters; these are principally the anharmonicity parameter and the dipolar-moment derivative. Varying reasonably the values of x_e and d', we have achieved an absolute coincidence between theoretical and experimental values for $\nu_{0\to 1}$ and $\nu_{0\to 2}$. As a result, the agreement occurs at values

$$x_e = 0.0188 \text{ and } d' = 0.219;$$

these specified parameters for SH ions, that form pairs in KCl, are hence sufficiently near those of isolated defects.

5 The Method of Factorization

Algebraic Formalism

To solve most equations of quantum mechanics, one generally applies the powerful apparatus of mathematical physics, which is based on traditional methods of the theory of integro-differential equations. Many problems might otherwise be solved in a purely algebraic manner. For instance, to describe a vibrational system in quantum mechanics, one uses the model of an anharmonic oscillator. This simple model might be regarded as being founded on an exact solution of the problem for a harmonic case that substantially represents the description of some physical system in an approximation of zero order. As the necessity to take into account the influence of anharmonicity increases, this solution becomes improved through the pertinent methods of perturbation theory. The non-zero matrix elements between the corresponding states of a perturbed system determine the observable quantities. In the case of the first few orders of the theory, the matrix elements are readily calculated in an algebraic manner, for instance, in the framework of a formalism of creation and destruction operators that follows from the classical works by Fock and Dirac. The calculations of higher orders are generally performed with the aid of special methods; the recurrence formalism of the perturbation theory in terms of the polynomials of quantum numbers might serve as one example of these special devices.

In some applications it is convenient to use the Morse oscillator instead of the harmonic oscillator as a zero-order approximation. Applying in this case the recurrence formalism of the perturbation theory, one might, in a manner analogous to the solution [16], evaluate the influence on the energy levels and the matrix elements of a term additional to the anharmonic field of the Morse potential that plays the role of a perturbation. Similar conclusions are applicable to another important case in which a non-perturbed system is described through the states in the form of the solutions of Schrödinger's equation for the Pöschl–Teller potential. Moreover, in a search for a solution of each such problem, the methods of non-commutative algebra according to the language of the so-called ladder operators [64], which are substantially the same of those as the creation and destruction operators, become applicable.

For a concrete physical problem, the terminology of ladder operators might be introduced in more than one way. For instance, a traditional analysis involves appropriate recurrence relations for special functions corresponding to the exact solutions of Schrödinger's equation, in which the well-known Morse- and Pöschl–Teller-type functions appear in the role of the potential and also a series of

Quantum Theory of Anharmonic Effects in Molecules. DOI: http://dx.doi.org/10.1016/B978-0-12-397912-4.00005-2

other simple potential functions [64]. The purely algebraic methods of factorization are applied less commonly. Among the latter algebraic enunciations is a technique of factorization described by Green [65] that is simple and elegant: we consider it in detail.

Suppose there exist in a set the q-numbers

η_1, η_2 and so on.

We determine an operator

$$F = \eta_1^+ \eta_1 + f_1 \equiv F_1,$$

in which f_1 is the physical number that has the maximum value possible for this representation. One puts, by definition,

$$F_2 = \eta_1 \eta_1^+ + f_1;$$

otherwise we suggest that

$$F_2 = \eta_2^+ \eta_2 + f_2,$$

in which $f_2 \geq f_1$. According to this scenario, for arbitrary positive integer n, we have

$$F_n = \eta_n^+ \eta_n + f_n$$

and

$$F_{n+1} = \eta_n \eta_n^+ + f_n.$$

Let us introduce a vector

$$|\varphi_n\rangle = \eta_n \eta_{n-1} \cdots \eta_1 |\psi\rangle,$$

in which $|\psi\rangle$ is some normalized eigenvector of operator F belonging to eigenvalue f. As

$$F|\psi\rangle = f|\psi\rangle \text{ and } \langle\psi|\psi\rangle = 1,$$

we have

$$\langle\varphi_1|\varphi_1\rangle = \langle\psi|\eta_1^+ \eta_1|\psi\rangle = \langle\psi|(F - f_1)|\psi\rangle = f - f_1,$$

where from, taking into account that

$$\langle\varphi_1|\varphi_1\rangle \geq 0,$$

there follows this inequality,

$$f \geq f_1.$$

Furthermore,

$$F_{n+1}\eta_n = \eta_n\eta_n^+\eta_n + \eta_n f_n = \eta_n F_n;$$

consequently,

$$\begin{aligned}
\langle\varphi_n|\varphi_n\rangle &= \langle\psi|\eta_1^+\eta_2^+\cdots\eta_n^+\eta_n\eta_{n-1}\cdots\eta_1|\psi\rangle \\
&= \langle\psi|\eta_1^+\eta_2^+\cdots(F_n - f_n)\eta_{n-1}\cdots\eta_1|\psi\rangle \\
&= \langle\psi|\eta_1^+\eta_2^+\cdots\eta_{n-1}^+\eta_{n-1}(F_{n-1} - f_n)\cdots\eta_1|\psi\rangle \\
&= \langle\psi|\eta_1^+\eta_2^+\cdots\eta_{n-1}^+\eta_{n-1}\cdots\eta_1(F_1 - f_n)|\psi\rangle \\
&= \langle\psi|\eta_1^+\eta_2^+\cdots(F_{n-1} - f_{n-1})\eta_{n-2}\cdots\eta_1|\psi\rangle(f - f_n) \\
&= \cdots = (f - f_1)(f - f_2)\cdots(f - f_n) \geq 0,
\end{aligned}$$

such that either

$$f \geq f_n$$

or

$$(f - f_1)(f - f_2)\cdots(f - f_{n-1}) = 0.$$

Quantity f is thus either more than each physical number $f_1, f_2, \ldots, f_n$ or equal to one of them. The obtained result is highly important: quantities $f_1, f_2, \ldots$, which are represented in order of increasing magnitude, constitute the eigenvalues of operator F.

We proceed to construct the eigenvectors of operator F. Let $|n-1\rangle$ be the eigenvector of F with eigenvalue f_n; one assumes that $|\psi\rangle = |n-1\rangle$. We have

$$\langle\varphi_{n-1}|\varphi_{n-1}\rangle > 0$$

and

$$\langle\varphi_n|\varphi_n\rangle = \langle n-1|\eta_1^+\eta_2^+\cdots\eta_{n-1}^+\eta_{n-1}\cdots\eta_1(F - f_n)|n-1\rangle = 0,$$

because $F|n-1\rangle = f_n|n-1\rangle$, hence

$$|\varphi_n\rangle = 0 \text{ or } \eta_n|\varphi_{n-1}\rangle = 0.$$

Furthermore,

$$(F_n - f_n)|\varphi_{n-1}\rangle = \eta_n^+\eta_n|\varphi_{n-1}\rangle = 0;$$

f_n is hence an eigenvalue of operator F_n with vector $|\varphi_{n\text{-}1}\rangle$. As

$$F_n\eta_n^+ = \eta_n^+\eta_n\eta_n^+ + \eta_n^+ f_n = \eta_n^+ F_{n+1}$$

and multiplying $\eta_1^+\eta_2^+\cdots\eta_{n-1}^+$ by F_n, we obtain

$$\eta_1^+\eta_2^+\cdots\eta_{n-1}^+F_n = \eta_1^+\eta_2^+\cdots F_{n-1}\eta_{n-1}^+ = F\eta_1^+\eta_2^+\cdots\eta_{n-1}^+$$

and

$$\eta_1^+\eta_2^+\cdots\eta_{n-1}^+F_n|\varphi_{n-1}\rangle = F\eta_1^+\eta_2^+\cdots\eta_{n-1}^+|\varphi_{n-1}\rangle = f_n\eta_1^+\eta_2^+\cdots\eta_{n-1}^+|\varphi_{n-1}\rangle.$$

Operating on the other side, $F|n-1\rangle = f_n|n-1\rangle$. This result is consequently accurate within a constant factor,

$$|n-1\rangle = \eta_1^+\eta_2^+\cdots\eta_{n-1}^+|\varphi_{n-1}\rangle,$$

in which vector $|\varphi_{n-1}\rangle$ is determined by the equation

$$\eta_n|\varphi_{n-1}\rangle = 0.$$

Atom of Hydrogen Type

Following Green [65], as an illustration to apply the algebraic method of factorization, we consider the calculation of the energy for an electron of an atom of hydrogen type with a Hamiltonian

$$H = \frac{p_r^2}{2m} + \frac{\hbar^2\ell(\ell+1)}{2mr^2} - \frac{Ze^2}{r},$$

in which p_r is the radial momentum, m is the reduced mass of the nucleus and the electron, ℓ is the orbital quantum number, r is the distance between the electron of charge $-e$ and the nucleus of charge Ze; for hydrogen, $Z = 1$. We assume

$$F = 2mH,$$

so

$$F = p_r^2 + \frac{\hbar^2\ell(\ell+1)}{r^2} - \frac{2\kappa}{r}, \quad \kappa = Ze^2m.$$

Let

$$\eta_n = p_r + \mathrm{i}\left(a_n + \frac{b_n}{r}\right),$$

in which a_n and b_n are real quantities, the explicit form of which one must find. We calculate $\eta_n^+\eta_n$ as

$$\eta_n^+\eta_n = (p_r - \mathrm{i}(a_n + b_n/r))(p_r + \mathrm{i}(a_n + b_n/r))$$
$$= p_r^2 + \mathrm{i}b_n[p_r, 1/r] + a_n^2 + \frac{2a_nb_n}{r} + \frac{b_n^2}{r^2} = p_r^2 + a_n^2 + \frac{2a_nb_n}{r} + \frac{b_n^2 - \hbar b_n}{r^2},$$

in which one takes into account that

$$[p_r, 1/r] = -\mathrm{i}\hbar\frac{\partial}{\partial r}(1/r) = \frac{\mathrm{i}\hbar}{r^2}.$$

In an analogous manner, one obtains

$$\eta_n\eta_n^+ = p_r^2 + a_n^2 + \frac{2a_nb_n}{r} + \frac{b_n^2 + \hbar b_n}{r^2}.$$

We define operator F_1:

$$F_1 = \eta_1^+\eta_1 + f_1 = p_r^2 + a_1^2 + \frac{2a_1b_1}{r} + \frac{b_1^2 - \hbar b_1}{r^2} + f_1.$$

On the other side,

$$F_1 = p_r^2 + \frac{\hbar^2\ell(\ell+1)}{r^2} - \frac{2\kappa}{r}.$$

On comparison, we obtain the equations

$$a_1b_1 = -\kappa,\ b_1(b_1 - \hbar) = \hbar^2\ell(\ell+1) \text{ and } a_1^2 + f_1 = 0.$$

Here, one might find two solutions. In the first case,

$$b_1 = -\hbar\ell,$$

then

$$a_1 = \frac{\kappa}{\hbar\ell} \text{ and } f_1 = -\frac{\kappa^2}{\hbar^2\ell^2}.$$

If

$$b_1 = \hbar(\ell+1),$$

then

$$a_1 = -\frac{\kappa}{\hbar(\ell + 1)} \text{ and } f_1 = -\frac{\kappa^2}{\hbar^2(\ell+1)^2}.$$

As

$$-\frac{\kappa^2}{\hbar^2(\ell+1)^2} > -\frac{\kappa^2}{\hbar^2\ell^2},$$

one chooses the second case.

We now compare the two expressions for F_{n+1},

$$\eta_n\eta_n^+ + f_n = \eta_{n+1}^+\eta_{n+1} + f_{n+1}$$

or in an explicit form

$$p_r^2 + a_n^2 + \frac{2a_nb_n}{r} + \frac{b_n^2 + \hbar b_n}{r^2} + f_n = p_r^2 + a_{n+1}^2 + \frac{2a_{n+1}b_{n+1}}{r} + \frac{b_{n+1}^2 - \hbar b_{n+1}}{r^2} + f_{n+1},$$

where from

$$a_nb_n = a_{n+1}b_{n+1},$$
$$b_n(b_n + \hbar) = b_{n+1}(b_{n+1} - \hbar)$$

and

$$a_n^2 + f_n = a_{n+1}^2 + f_{n+1}.$$

If

$$b_n = -b_{n+1}, \text{ then } a_n = -a_{n+1} \text{ and } f_n = f_{n+1},$$

this case fails to have physical interest.

Let us consider the second possibility when

$$b_{n+1} = b_n + \hbar.$$

We have

$$b_{n+1} = b_n + \hbar = b_{n-1} + 2\hbar = \cdots = b_1 + n\hbar = \hbar(\ell + n + 1);$$

$$a_{n+1}b_{n+1} = a_nb_n = a_{n-1}b_{n-1} = \cdots = a_1b_1 = -\kappa,$$

where from

$$a_n = -\frac{\kappa}{b_n} = -\frac{\kappa}{\hbar(\ell + n)}.$$

Taking into account that

$$a_{n+1}^2 + f_{n+1} = a_n^2 + f_n = \cdots = a_1^2 + f_1 = 0,$$

we find the eigenvalues of quantity F:

$$f_n = -a_n^2 = -\frac{\kappa^2}{\hbar^2(\ell+n)^2}.$$

Eigenvalues E of Hamiltonian H are thus

$$E = \frac{f_n}{2m} = -\frac{\kappa^2}{2m\hbar^2(\ell+n)^2}$$

or

$$E_\nu = -\frac{mZ^2e^4}{2\hbar^2\nu^2}, \quad \nu = 1, 2, \ldots.$$

As the set of c-numbers E_1, E_2 and so on is restricted by value $E = 0$ from above, then, according to this inequality,

$$(E - E_1)(E - E_2)\cdots(E - E_n) \geq 0$$

of Green's formalism, quantity E must be either equal to one value E_1, E_2,. . ., or equal to any value from zero until infinity. For $E < 0$, the energy levels constitute a discrete spectrum, for which the electron is in a bound state. For $E > 0$, there is no bound state for the electron; the energy spectrum is continuous.

Some Problems Involving Anharmonicity

Applying the algebraic method of factorization, we solve problems for eigenvalues of important Hamiltonians in a series that describe simple anharmonic systems. So, let

$$H = \frac{p_r^2}{2m} + V_r$$

generally be the Hamiltonian of some physical system, which is a particle that moves in a given anharmonic potential V_r; $V_r = V(x)$. Here, r is the current coordinate of a particle of momentum p_r and mass m;

$$x = \frac{r - r_0}{r_0}$$

is the relative shift of coordinate r from its equilibrium value r_0. The scheme to determine the eigenvalues is simple. First, we postulate that

$$F = 2mH$$

and, according to insight, we choose variable η_n. Then, on comparing two expressions for F_1, we find f_1. To value f_1 corresponds the state of the system with least energy E_0; $f_1 = 2mE_0$. Other quantities $f_2, f_3, \ldots$ follow from a comparison of two expressions for F_{n+1}. As

$$F|n\rangle = f_{n+1}|n\rangle,$$

the sought eigenvalues E_n, corresponding to eigenstates $|n\rangle$, are expressible through the formula

$$E_n = \frac{f_{n+1}}{2m}.$$

This scenario to find a solution is general; note that our interest is focused on the energy levels of bound states – we ignore a continuous spectrum of energy. We proceed to consider some examples.

Pöschl–Teller Potential

For a particle of mass m, the Hamiltonian

$$H = \frac{p_r^2}{2m} - \frac{D}{\cosh^2(\alpha x)}$$

contains

$$V(x) = -\frac{D}{\cosh^2(\alpha x)},$$

which is called a modified Pöschl–Teller potential; D and α are adjustable parameters. According to the general scenario, we put

$$F = p_r^2 - \frac{A}{\cosh^2(\alpha x)}, \quad A = 2mD.$$

With variable η_n chosen in a form

$$\eta_n = p_r + \mathrm{i} b_n \tanh(\alpha x),$$

we calculate coefficient b_n and the eigenvalues of H.

Taking into account that

$$[\tanh(\alpha x), p_r] = \mathrm{i}\hbar \frac{1}{r_0}\frac{\partial}{\partial x}\tanh(\alpha x) = \frac{\mathrm{i}\hbar\alpha}{r_0 \cosh^2(\alpha x)},$$

and also

$$\tanh^2(\alpha x) = 1 - \frac{1}{\cosh^2(\alpha x)},$$

we have

$$\begin{aligned}\eta_n^+ \eta_n &= p_r^2 - \mathrm{i} b_n [\tanh(\alpha x), p_r] + b_n^2 \tanh^2(\alpha x) \\ &= p_r^2 + b_n^2 + \left(\frac{\alpha\hbar}{r_0} b_n - b_n^2\right)\frac{1}{\cosh^2(\alpha x)}\end{aligned}$$

and

$$\eta_n \eta_n^+ = p_r^2 + b_n^2 - \left(\frac{\alpha\hbar}{r_0} b_n + b_n^2\right)\frac{1}{\cosh^2(\alpha x)}.$$

We compare two expressions for F_1; on one side

$$F_1 = p_r^2 - \frac{A}{\cosh^2(\alpha x)},$$

and on the other side

$$F_1 = \eta_1^+ \eta_1 + f_1 = p_r^2 + b_1^2 + \left(\frac{\alpha\hbar}{r_0} b_1 - b_1^2\right)\frac{1}{\cosh^2(\alpha x)} + f_1;$$

consequently,

$$\frac{\alpha\hbar}{r_0} b_1 - b_1^2 = -A \text{ and } b_1^2 + f_1 = 0.$$

On solving these equations, we find that

$$b_1 = \frac{\alpha\hbar}{2r_0} \pm \sqrt{A + \frac{\alpha^2\hbar^2}{4r_0^2}} \text{ and } f_1 = -b_1^2.$$

The maximum value for f_1 occurs for

$$b_1 = \frac{\alpha\hbar}{2r_0} - \sqrt{A + \frac{\alpha^2\hbar^2}{4r_0^2}},$$

hence

$$f_1 = -\left(\frac{\alpha\hbar}{2r_0} - \sqrt{A + \frac{\alpha^2\hbar^2}{4r_0^2}}\right)^2.$$

We compare two expressions for F_{n+1}. We have

$$\eta_n\overset{+}{\eta_n} + f_n = \overset{+}{\eta}_{n+1}\eta_{n+1} + f_{n+1},$$

i.e.

$$p_r^2 + b_n^2 - \left(\frac{\alpha\hbar}{r_0}b_n + b_n^2\right)\frac{1}{\cosh^2(\alpha x)} + f_n$$
$$= p_r^2 + b_{n+1}^2 + \left(\frac{\alpha\hbar}{r_0}b_{n+1} - b_{n+1}^2\right)\frac{1}{\cosh^2(\alpha x)} + f_{n+1},$$

where from

$$b_{n+1}\left(b_{n+1} - \frac{\alpha\hbar}{r_0}\right) = b_n\left(b_n + \frac{\alpha\hbar}{r_0}\right)$$

and

$$b_{n+1}^2 + f_{n+1} = b_n^2 + f_n = \cdots = b_1^2 + f_1 = 0.$$

From the former relation we obtain that either $b_{n+1} = -b_n$ or

$$b_{n+1} = b_n + \frac{\alpha\hbar}{r_0},$$

of which the former solution is inappropriate. In the latter case we have

$$b_{n+1} = b_n + \frac{\alpha\hbar}{r_0} = b_{n-1} + \frac{2\alpha\hbar}{r_0} = \cdots = b_1 + \frac{n\alpha\hbar}{r_0};$$

as a result,

$$f_{n+1} = -b_{n+1}^2 = -\frac{\alpha^2\hbar^2}{r_0^2}\left[\sqrt{\frac{A}{\alpha^2\hbar^2}r_0^2 + \frac{1}{4}} - \frac{1}{2} - n\right]^2.$$

The sought energy levels E_n of a particle that moves in a field according to the modified Pöschl–Teller potential become thus determined through the formula

$$E_n = \frac{f_{n+1}}{2m} = -\frac{\alpha^2\hbar^2}{2mr_0^2}\left[\sqrt{\frac{2mD}{\alpha^2\hbar^2}r_0^2 + \frac{1}{4}} - \frac{1}{2} - n\right]^2.$$

Pöschl–Teller-Like Potential

Another pertinent instance is the motion of a particle in a field with the potential

$$V(x) = V_0 \tan^2\left(\frac{\pi x}{L}\right),$$

in which V_0 and L are parameters, $x \in [-L/2, L/2]$; this potential belongs to the Pöschl–Teller type. In this case the Hamiltonian has a form

$$H = \frac{p_r^2}{2m} + V_0 \tan^2\left(\frac{\pi x}{L}\right).$$

We suppose that

$$F = p_r^2 + A \tan^2(\alpha x)$$

and

$$\eta_n = p_r - \mathrm{i}a_n \tan(\alpha x).$$

Here, $A = 2mV_0$ and $\alpha = \pi/L$; coefficient a_n remains to be determined.

Taking into account that

$$[\tan(\alpha x), p_r] = \mathrm{i}\hbar\alpha \cos^{-2}(\alpha x) = \mathrm{i}\hbar\alpha(1 + \tan^2(\alpha x)),$$

we calculate $\eta_n^+\eta_n$:

$$\eta_n^+\eta_n = p_r^2 - \hbar\alpha a_n + (a_n^2 - \hbar\alpha a_n)\tan^2(\alpha x);$$

in an analogous manner, we find that

$$\eta_n\eta_n^+ = p_r^2 + \hbar\alpha a_n + (a_n^2 + \hbar\alpha a_n)\tan^2(\alpha x).$$

For operator F_1, from one side,

$$F_1 = \eta_1^+ \eta_1 + f_1 = p_r^2 - \hbar\alpha a_1 + (a_1^2 - \hbar\alpha a_1)\tan^2(\alpha x) + f_1;$$

from the other side,

$$F_1 = p_r^2 + A\tan^2(\alpha x).$$

On comparing, we obtain these two equations

$$f_1 - \hbar\alpha a_1 = 0 \text{ and } a_1^2 - \hbar\alpha a_1 = A,$$

where from

$$f_1 = \hbar\alpha a_1 \text{ and } a_1 = \frac{\hbar\alpha}{2} \pm \sqrt{A + \frac{(\hbar\alpha)^2}{4}};$$

for the maximum value for f_1,

$$a_1 = \frac{\hbar\alpha}{2} + \sqrt{A + \frac{(\hbar\alpha)^2}{4}},$$

we hence choose exactly this value for a_1.

Furthermore, uncovering this identity

$$\eta_n \eta_n^+ + f_n = \eta_{n+1}^+ \eta_{n+1} + f_{n+1},$$

we obtain the following equations:

$$\hbar\alpha a_n + f_n = -\hbar\alpha a_{n+1} + f_{n+1} \text{ and } a_n(a_n + \hbar\alpha) = a_{n+1}(a_{n+1} - \hbar\alpha).$$

Solution $a_{n+1} = -a_n$ has no physical meaning, hence

$$a_{n+1} = a_n + \hbar\alpha = \cdots = a_1 + n\hbar\alpha.$$

To determine f_{n+1}, we write

$$\begin{aligned} f_{n+1} - \hbar\alpha a_{n+1} &= f_n + \hbar\alpha a_n, \\ f_n - \hbar\alpha a_n &= f_{n-1} + \hbar\alpha a_{n-1}, \\ &\vdots \\ f_2 - \hbar\alpha a_2 &= f_1 + \hbar\alpha a_1; \end{aligned}$$

on summing these equations, we find

$$\begin{aligned} f_{n+1} &= f_1 + \hbar\alpha(a_1 + a_2 + \cdots + a_n) + \hbar\alpha(a_2 + \cdots + a_n + a_{n+1}) \\ &= f_1 + \hbar\alpha(2(a_1 + a_2 + \cdots + a_n) + n\hbar\alpha). \end{aligned}$$

As $f_1 = \hbar\alpha a_1$ and

$$a_1 + a_2 + \cdots + a_n = na_1 + \hbar\alpha(1 + 2 + \cdots + (n-1)) = na_1 + \frac{\hbar\alpha}{2}n(n-1),$$

we have

$$\begin{aligned} f_{n+1} &= (\hbar\alpha n)^2 + 2\hbar\alpha a_1 n + \hbar\alpha a_1 = (\hbar\alpha)^2(n + a_1/\hbar\alpha)^2 - a_1^2 + \hbar\alpha a_1 \\ &= (\hbar\alpha)^2\left(n + \frac{a_1}{\hbar\alpha}\right)^2 - A. \end{aligned}$$

The eigenvalues of Hamiltonian H are consequently equal to

$$E_n = \frac{f_{n+1}}{2m} = \frac{1}{2m}\left(\frac{\hbar\pi}{L}\right)^2\left(n + \frac{1}{2} + \frac{1}{2}\sqrt{\frac{8mV_0}{\hbar^2\pi^2}L^2 + 1}\right)^2 - V_0.$$

The problem is solved.

Morse's Oscillator

We show that the developed theory generates the correct values for the energy levels of Morse's oscillator. We write the Hamiltonian in a form

$$H = \frac{p_r^2}{2m} + D(1 - \mathrm{e}^{-a_M x})^2;$$

we assume

$$F = p_r^2 + A - 2A\,\mathrm{e}^{-a_M x} + A\,\mathrm{e}^{-2a_M x}$$

and

$$\eta_n = p_r + \mathrm{i}(b_n + c_n\,\mathrm{e}^{-a_M x}).$$

Here, D and a_M are the parameters of Morse's potential, $A = 2mD$; b_n and c_n are the real quantities, the explicit forms of which are to be defined.

We calculate $\eta_n^+\eta_n$ as

$$\eta_n^+\eta_n = p_r^2 - \mathrm{i}c_n[\mathrm{e}^{-a_M x}, p_r] + (b_n + c_n\,\mathrm{e}^{-a_M x})^2$$
$$= p_r^2 + (2b_n - \hbar a_M/r_0)c_n\,\mathrm{e}^{-a_M x} + b_n^2 + c_n^2\,\mathrm{e}^{-2a_M x},$$

in which we take into account that

$$[\mathrm{e}^{-a_M x}, p_r] = \mathrm{i}\hbar\frac{\partial}{\partial r}\mathrm{e}^{-a_M x} = -\,\mathrm{i}\hbar\frac{a_M}{r_0}\mathrm{e}^{-a_M x}.$$

In an analogous manner, one finds

$$\eta_n\eta_n^+ = p_r^2 + (2b_n + \hbar a_M/r_0)c_n\,\mathrm{e}^{-a_M x} + b_n^2 + c_n^2\,\mathrm{e}^{-2a_M x}.$$

Let us consider operator F_1:

$$F_1 = \eta_1^+\eta_1 + f_1 = p_r^2 + (2b_1 - \hbar a_M/r_0)c_1\,\mathrm{e}^{-a_M x} + b_1^2 + c_1^2\,\mathrm{e}^{-2a_M x} + f_1.$$

From the other side,

$$F_1 = p_r^2 + A - 2A\,\mathrm{e}^{-a_M x} + A\,\mathrm{e}^{-2a_M x}.$$

On comparing, we obtain the following equations:

$$b_1 = \frac{\hbar a_M}{2r_0} - \frac{A}{c_1}, \quad c_1^2 = A \text{ and } b_1^2 + f_1 = A.$$

Here, one might have two possible solutions. If $c_1 = -\sqrt{A}$, then

$$b_1 = \frac{\hbar a_M}{2r_0} + \sqrt{A} \text{ and } f_1 = A - \left(\frac{\hbar a_M}{2r_0} + \sqrt{A}\right)^2,$$

whereas for $c_1 = \sqrt{A}$,

$$b_1 = \frac{\hbar a_M}{2r_0} - \sqrt{A} \text{ and } f_1 = A - \left(\frac{\hbar a_M}{2r_0} - \sqrt{A}\right)^2.$$

Because, in the latter case, the value of quantity f_1 is greater, one chooses the second solution:

$$f_1 = A - \left(\frac{\hbar a_M}{2r_0} - \sqrt{A}\right)^2 = \frac{\hbar a_M}{r_0}\sqrt{A} - \left(\frac{\hbar a_M}{2r_0}\right)^2.$$

As $\hbar\omega = 2Da_M^2\lambda^2$, in which ω is the vibrational frequency, and $\lambda = (1/r_0)\sqrt{\hbar/m\omega}$, then

$$a_M = \frac{1}{\lambda}\sqrt{\frac{\hbar\omega}{2D}} = r_0\omega\sqrt{\frac{m}{2D}}.$$

Consequently,

$$f_1 = m\hbar\omega - m\frac{(\hbar\omega)^2}{8D},$$

and the least eigenvalue of Hamiltonian H equals

$$E_0 = \frac{f_1}{2m} = \hbar\omega\left(0 + \frac{1}{2}\right) - \frac{(\hbar\omega)^2}{4D}\left(0 + \frac{1}{2}\right)^2.$$

We find other eigenvalues from a comparison of the two expressions for F_{n+1}. By definition,

$$\eta_n\eta_n^+ + f_n = \eta_{n+1}^+\eta_{n+1} + f_{n+1},$$

i.e.

$$\begin{aligned} &p_r^2 + (2b_n + \hbar a_M/r_0)c_n\,\mathrm{e}^{-a_M x} + b_n^2 + c_n^2\,\mathrm{e}^{-2a_M x} + f_n \\ &= p_r^2 + (2b_{n+1} - \hbar a_M/r_0)c_{n+1}\,\mathrm{e}^{-a_M x} + b_{n+1}^2 + c_{n+1}^2\,\mathrm{e}^{-2a_M x} + f_{n+1}; \end{aligned}$$

where from

$$\begin{aligned} &c_{n+1}^2 = c_n^2, \\ &(2b_{n+1} - \hbar a_M/r_0)c_{n+1} = (2b_n + \hbar a_M/r_0)c_n, \\ &b_{n+1}^2 + f_{n+1} = b_n^2 + f_n. \end{aligned}$$

We see that

$$c_{n+1}^2 = c_n^2 = c_{n-1}^2 = \cdots = c_1^2 = A;$$

discarding the solution $c_n = -\sqrt{A}$, one obtains

$$c_n = \sqrt{A}.$$

Furthermore,

$$b_{n+1} = b_n + \frac{\hbar a_M}{r_0} = b_{n-1} + 2\frac{\hbar a_M}{r_0} = \cdots = b_1 + n\frac{\hbar a_M}{r_0} = -\sqrt{A} + \left(n + \frac{1}{2}\right)\frac{\hbar a_M}{r_0}.$$

Finally,

$$b_{n+1}^2 + f_{n+1} = b_n^2 + f_n = \cdots = b_1^2 + f_1 = A,$$

where from

$$f_{n+1} = A - b_{n+1}^2 = 2\frac{\hbar a_M}{r_0}\sqrt{A}\left(n+\frac{1}{2}\right) - (\hbar a_M/r_0)^2\left(n+\frac{1}{2}\right)^2.$$

Taking into account that

$$2\frac{\hbar a_M}{r_0}\sqrt{A} = 2m\hbar\omega \text{ and } (\hbar a_M/r_0)^2 = 2m\frac{(\hbar\omega)^2}{4D},$$

we have

$$f_{n+1} = 2m\hbar\omega\left(n+\frac{1}{2}\right) - 2m\frac{(\hbar\omega)^2}{4D}\left(n+\frac{1}{2}\right)^2.$$

As $E_n = f_{n+1}/2m$,

$$E_n = \hbar\omega\left(n+\frac{1}{2}\right) - \frac{(\hbar\omega)^2}{4D}\left(n+\frac{1}{2}\right)^2,$$

which is the required solution.

Generalized Morse's Oscillator

As a further and more complicated instance, we consider this potential,

$$V(x) = D\left(\frac{1-\mathrm{e}^{-ax}}{1-k\,\mathrm{e}^{-ax}}\right)^2 = D\left(1 + 2\frac{k-1}{\mathrm{e}^{ax}-k} + \frac{(k-1)^2}{(\mathrm{e}^{ax}-k)^2}\right),$$

in which $a = (1-k)a_M$ and $|k| < 1$ [66]; a_M, D and k are adjustable parameters. The corresponding Hamiltonian has a form

$$H = \frac{p_r^2}{2m} + D\left(1 + 2\frac{k-1}{\mathrm{e}^{ax}-k} + \frac{(k-1)^2}{(\mathrm{e}^{ax}-k)^2}\right).$$

We define quantity F as $F = 2m(H - D)$, i.e.

$$F = p_r^2 + \frac{A}{y} + \frac{B}{y^2},$$

in which $y = e^{ax} - k$, $A = 4mD(k-1)$ and $B = 2mD(k-1)^2$. Moreover, we put

$$\eta_n = p_r + i\left(b_n + \frac{c_n}{e^{ax} - k}\right);$$

with $k = 0$ this variable transforms into analogous quantity η_n for the case of Morse's oscillator; we must determine coefficients b_n and c_n.

According to our scenario, we begin from the calculation of $\eta_n^+ \eta_n$:

$$\eta_n^+ \eta_n = p_r^2 - ic_n[1/y, p_r] + b_n^2 + \frac{2b_n c_n}{y} + \frac{c_n^2}{y^2},$$

as

$$[1/y, p_r] = -i\hbar \frac{a}{r_0}\left(\frac{1}{y} + \frac{k}{y^2}\right),$$

then

$$\eta_n^+ \eta_n = p_r^2 + b_n^2 + \frac{1}{y}\left(2b_n c_n - \frac{\hbar a}{r_0} c_n\right) + \frac{1}{y^2}\left(c_n^2 - \frac{k\hbar a}{r_0} c_n\right);$$

analogously we find that

$$\eta_n \eta_n^+ = p_r^2 + b_n^2 + \frac{1}{y}\left(2b_n c_n + \frac{\hbar a}{r_0} c_n\right) + \frac{1}{y^2}\left(c_n^2 + \frac{k\hbar a}{r_0} c_n\right).$$

Let us consider quantity F_1:

$$F_1 = \eta_1^+ \eta_1 + f_1 = p_r^2 + b_1^2 + \frac{1}{y}\left(2b_1 c_1 - \frac{\hbar a}{r_0} c_1\right) + \frac{1}{y^2}\left(c_1^2 - \frac{k\hbar a}{r_0} c_1\right) + f_1$$
$$\equiv p_r^2 + \frac{A}{y} + \frac{B}{y^2},$$

where from

$$b_1^2 + f_1 = 0,\ 2b_1 c_1 - \frac{\hbar a}{r_0} c_1 = A \text{ and } c_1^2 - \frac{k\hbar a}{r_0} c_1 = B.$$

On solving the latter equation with respect to c_1, we obtain

$$c_1 = \frac{k\hbar a}{2r_0} \pm \sqrt{B + \left(\frac{k\hbar a}{2r_0}\right)^2};$$

consequently,

$$b_1 = \frac{A}{(k\hbar a/r_0) \pm 2\sqrt{B + (k\hbar a/2r_0)^2}} + \frac{\hbar a}{2r_0}.$$

Choosing b_1 that leads to the maximum value for f_1, we have

$$f_1 = -b_1^2 = -\left(\frac{\hbar a}{2r_0} + A\left(\frac{k\hbar a}{r_0} + 2\sqrt{B + \left(\frac{k\hbar a}{2r_0}\right)^2}\right)^{-1}\right)^2.$$

To calculate other quantities f_n, we consider the identity

$$\eta_n \eta_n^+ + f_n = \eta_{n+1}^+ \eta_{n+1} + f_{n+1}$$

or, in an explicit form,

$$p_r^2 + b_n^2 + \frac{1}{y}\left(2b_n c_n + \frac{\hbar a}{r_0} c_n\right) + \frac{1}{y^2}\left(c_n^2 + \frac{k\hbar a}{r_0} c_n\right) + f_n$$
$$= p_r^2 + b_{n+1}^2 + \frac{1}{y}\left(2b_{n+1} c_{n+1} - \frac{\hbar a}{r_0} c_{n+1}\right) + \frac{1}{y^2}\left(c_{n+1}^2 - \frac{k\hbar a}{r_0} c_{n+1}\right) + f_{n+1}.$$

On comparing the left and right parts of this identity, we find these equations

$$b_{n+1}^2 + f_{n+1} = b_n^2 + f_n = \cdots = b_1^2 + f_1 = 0,$$

$$2b_{n+1} c_{n+1} - \frac{\hbar a}{r_0} c_{n+1} = 2b_n c_n + \frac{\hbar a}{r_0} c_n \text{ and } c_{n+1}\left(c_{n+1} - \frac{k\hbar a}{r_0}\right) = c_n\left(c_n + \frac{k\hbar a}{r_0}\right).$$

From the latter relation, discarding the solution $c_{n+1} = -c_n$, we obtain

$$c_{n+1} = c_n + \frac{k\hbar a}{r_0} = \cdots = c_1 + n\frac{k\hbar a}{r_0}.$$

In turn, we determine b_{n+1}, having summed all equations of this system

$$2b_{n+1} c_{n+1} - \gamma c_{n+1} = 2b_n c_n + \gamma c_n,$$
$$2b_n c_n - \gamma c_n = 2b_{n-1} c_{n-1} + \gamma c_{n-1},$$
$$\vdots$$
$$2b_2 c_2 - \gamma c_2 = 2b_1 c_1 + \gamma c_1,$$

in which $\gamma = \hbar a/r_0$. As a result,

$$2b_{n+1}c_{n+1} = 2b_1c_1 + \gamma(c_2 + \cdots + c_n + c_{n+1}) + \gamma(c_1 + \cdots + c_{n-1} + c_n);$$

as

$$2b_1c_1 - \gamma c_1 = A$$

and

$$c_1 + \cdots + c_{n-1} + c_n = nc_1 + k\gamma\frac{n(n-1)}{2},$$

then

$$2b_{n+1}c_{n+1} = A + \gamma c_{n+1} + 2\gamma n\left(c_1 + k\gamma\frac{n-1}{2}\right),$$

where from

$$b_{n+1} = \frac{A + 2\gamma n(c_1 + k\gamma((n-1)/2))}{2(c_1 + nk\gamma)} + \frac{\gamma}{2} = \frac{A - k\gamma^2 n(n+1)}{2(c_1 + nk\gamma)} + \gamma\left(n + \frac{1}{2}\right).$$

Supposing

$$n' = n + \frac{1}{2},\ n(n+1) = n'^2 - \frac{1}{4} \text{ and } c_1 + nk\gamma = k\gamma n' + \sqrt{B + \left(\frac{k\gamma}{2}\right)^2},$$

we simplify the obtained expression for b_{n+1}. We have

$$\begin{aligned} b_{n+1} &= \frac{A - k\gamma^2 n'^2 + (k\gamma^2)/4}{2k\gamma\left(n' + \operatorname{sgn}(k)\sqrt{B/(k\gamma)^2 + (1/4)}\right)} + \gamma n' \\ &\equiv \frac{(A/k\gamma) + (\gamma/4) - \gamma((n'+Q)-Q)^2}{2(n'+Q)} + \gamma n', \end{aligned}$$

in which $Q = \operatorname{sgn}(k)\sqrt{B/(k\gamma)^2 + (1/4)}$, hence

$$b_{n+1} = \frac{(A/k\gamma) + (\gamma/4) - \gamma Q^2}{2(n'+Q)} + \frac{\gamma}{2}(n'+Q).$$

Finally, taking into account that

$$\frac{A}{k\gamma} + \frac{\gamma}{4} - \gamma Q^2 = \frac{2mD}{\gamma}\left(1 - \frac{1}{k^2}\right),$$

we obtain

$$f_{n+1} = -b_{n+1}^2 = -mD(1 - k^{-2}) - \frac{m^2D^2}{\gamma^2} \cdot \frac{(1-k^{-2})^2}{(n'+Q)^2} - \frac{\gamma^2}{4}(n'+Q)^2.$$

The eigenvalues of Hamiltonian H are thus equal to

$$\frac{f_{n+1}}{2m} + D = \frac{D}{2}(1 + k^{-2}) - \frac{mD^2}{2\gamma^2} \cdot \frac{(1-k^{-2})^2}{(n'+Q)^2} - \frac{\gamma^2}{8m}(n'+Q)^2,$$

and

$$E_n = \frac{D}{2}\left[1 + k^{-2} - \frac{mDr_0^2}{\hbar^2 a^2} \cdot \frac{(1-k^{-2})^2}{(n+1/2+Q)^2} - \frac{\hbar^2 a^2}{4mDr_0^2}(n+1/2+Q)^2\right]$$

are the sought energy levels of the generalized Morse's oscillator;

$$Q = \mathrm{sgn}(k)\sqrt{\frac{2mDr_0^2}{\hbar^2 a^2}(1-k^{-1})^2 + \frac{1}{4}}.$$

References

1. Dunham JL: Intensities of vibration–rotation bands with special reference to those of HCl, *Phys Rev* 35: 1347–1354, 1930.
2. Hirschfelder JO, Byers Brown W, Epstein ST: Recent developments in perturbation theory, *Adv Quant Chem* 1: 255–385, 1964.
3. Amat G, Nielsen HH, Tarrago G: *Rotation–vibration of polyatomic molecules*, New York, 1971, Marcel Dekker Inc.
4. Kiselev AA: Adiabatic perturbation theory in molecular spectroscopy, *Can J Phys* 56: 615–647, 1978.
5. Papousek D, Aliev MR: *Molecular vibrational–rotational spectra*, Amsterdam, 1982, Elsevier.
6. Ogilvie JF: *The vibrational and rotational spectrometry of diatomic molecules*, London, 1998, Academic Press.
7. Hirschfelder JO: Classical and quantum-mechanical hypervirial theorems, *J Chem Phys* 33: 1462–1466, 1960.
8. Kiselev AA, Liapzev AV: *The quantum-mechanical perturbation theory (diagrammatic method)*, Leningrad, 1989, Leningrad State University.
9. Nielsen HH: The vibration–rotation energies of molecules, *Rev Mod Phys* 23: 90–136, 1951.
10. Tipping RH: Accurate analytic expectation values for an anharmonic oscillator using the hypervirial theorem, *J Chem Phys* 59: 6433–6442, 1973.
11. Tipping RH: Accurate analytic matrix elements for anharmonic oscillator using quantum mechanical commutator relations and sum rules, *J Chem Phys* 59: 6443–6449, 1973.
12. Geerlings P, Berckmans D, Figeys HP: The influence of electrical and mechanical anharmonicity on the vibrational transition moments of diatomic and polyatomic molecules, *J Mol Struct* 57: 283–297, 1979.
13. Aliev MR, Watson JKG: Higher-order effects in the vibration-rotation spectra of semirigid molecules. In Rao KN, editor: *Molecular spectroscopy: modern research*, New York, 1985, Academic Press, pp 1–67.
14. Sarka K, Demaison J: Perturbation theory, effective Hamiltonians and force constants. In Jensen P, Bunker PR, editors: *Computational molecular spectroscopy*, New York, 2000, Wiley, pp 255–303.
15. Watson JKG: Determination of centrifugal distortion coefficients of asymmetric-top molecules, *J Chem Phys* 46: 1935–1949, 1967.
16. Kazakov KV: Electro-optics of molecules, *Opt Spectrosc* 97: 725–734, 2004.
17. Kirzhnits DA: Formulation of quantum theory based on differentiation with respect to coupling parameter, *Problems of theoretical physics: a volume dedicated to the memory of Igor E. Tamm*, Moscow, 1972, Nauka.
18. Kazakov KV: Electro-optics of molecules. II, *Opt Spectrosc* 104: 477–490, 2008.
19. Kazakov KV: Doctoral Dissertation, 2006, St. Petersburg.
20. Kazakov KV: Formalism of quantum number polynomials, *Russ Phys J* 48: 954–965, 2005.

21. Watson JKG: Simplification of the molecular vibration–rotation Hamiltonian, *Mol Phys* 15: 479–490, 1968.
22. Bunker PR, Jensen P: *Molecular symmetry and spectroscopy*, Ottawa, 1998, NRC Research Press.
23. Smith MAH, Rinsland CP, Fridovich B, Rao KN: Intensities and collision broadening parameters from infrared spectra. In Rao KN, editor: *Molecular spectroscopy: modern research*, New York, 1985, Academic Press, pp 111–248.
24. Herman R, Wallis RF: Influence of vibration–rotational interaction on line intensities in the vibration–rotational bands of diatomic molecules, *J Chem Phys* 23: 637–646, 1955.
25. Gell-Mann M, Low F: Bound state in quantum field theory, *Phys Rev* 84: 350–354, 1951.
26. Brueckner KA: Many-body problem for strongly interacting particles. II. Linked cluster expansion, *Phys Rev* 100: 36–45, 1955.
27. Bethe HA: Nuclear many-body problem, *Phys Rev* 103: 1353–1390, 1956.
28. Goldstone J: Derivation of the Brueckner many-body theory, *Proc R Soc London, Ser A* 239: 267–279, 1957.
29. March NH, Young WH, Sampanthar S: *The many-body problem in quantum mechanics*, Cambridge, 1967, Cambridge University Press.
30. Hubbard J: The description of collective motions in terms of many-body perturbation theory, *Proc R Soc London, Ser A* 240: 539–560, 1957.
31. Kirzhnits DA: *Field theoretical methods in many-body systems, Oxford*, 1967, Pergamon.
32. Kittel C: *Quantum theory of solids*, New York, 1963, Wiley.
33. Mattuck RD: *A guide to Feynman diagrams in the many-body problem*, London, 1967, McGraw-Hill.
34. Tsvelik AM: *Quantum field theory in condensed matter physics*, Cambridge, 1998, Cambridge University Press.
35. Hellmann H: *Introduction to quantum chemistry*, Leipzig, 1937, Deuticke.
36. Feynman RP: Forces in molecules, *Phys Rev* 56: 340–343, 1939.
37. Foldy LL, Wouthuysen SA: On the Dirac theory of spin ½ particles and its non-relativistic limit, *Phys Rev* 78: 29–36, 1950.
38. Hirschfelder JO, Curtiss CF, Bird RB: *Molecular theory of gases and liquids*, New York, 1964, Wiley.
39. Makarewicz J: Renormalized perturbation theory for a general D-dimensional isotropic anharmonic oscillator, *J Phys A: Math Gen* 17: 1449–1460, 1984.
40. Wigner EP: *Group theory and its application to the quantum mechanics of atomic spectra*, New York, 1959, Academic Press.
41. Afanasiev AD, An CP, Luty F: IR anharmonicity study of the OH^- and OD^- stretch-mode in alkali halides. In Kanert O, Spaeth J-M, editors: *Proceedings of the XII international conference on defects in insulating materials (ICDIM-92)*, Nordkirchen, 1993, World Scientific, pp 551–554.
42. Woll AR, Fowler WB: XH defects in nonmetallic solids: general properties of Morse oscillators, *Phys Rev B* 48: 16788–16792, 1993.
43. Afanasiev AD, Ivanov AA, Luty F: FTIR stretching-mode measurements and calculations of equal and unequal pairs of SH^- and SD^- defects in KCl, *Radiation Effects & Defects in Solids* 155: 345–348, 2001.
44. Kazakov KV, Afanasiev AD: Overtone transitions of diatomic dimers XH^-–XH^- in ionic crystals, *Opt Spectrosc* 95: 54–59, 2003.

45. Kolomiitsova TD, Burtsev AP, Fedoseev VG, Shchepkin DN: Manifestation of interaction of the transition dipole moments in IR spectra of low-temperature liquids and solutions in liquefied noble gases, *Chem Phys* 238: 315–327, 1998.
46. Bulanin MO, Domanskaya AV, Kerl K: High-resolution FTIR measurement of the line parameters in the fundamental band of HI, *J Mol Spectrosc* 218: 75–79, 2003.
47. Bulanin MO, Domanskaya AV, Grigorev IM, Kerl K: Spectral line parameters in the $(2\leftarrow0)$ overtone band and the dipole moment function of the HI molecule, *J Mol Spectrosc* 223: 67–72, 2004.
48. Niay P, Bernage P, Coquant C, Houdart R: A measurement of the intensities of the vibration–rotational bands $0\rightarrow4$ and $0\rightarrow5$ of HI, *Can J Phys* 56: 727–736, 1978.
49. Mills IM: Harmonic and anharmonic force field calculations. In Dixon RN, editor: *Theoretical chemistry, vol I, Specialist periodical reports of the Chemical Society*, London, 1974, The Chemical Society, pp 200–235.
50. Kazakov KV, Gorbacheva MA: Calculation of higher-order approximations of the coefficients of the Herman–Wallis factor. Test for hydrogen halides, *Opt Spectrosc* 106: 475–482, 2009.
51. Van Stralen JNP, Visscher L, Ogilvie JF: Theoretical and experimental evaluation of the radial function for electric dipole moment of hydrogen iodide, *Phys Chem Chem Phys* 6: 3779–3785, 2004.
52. Guelachvili G, Niay P, Bernage P: Fourier transform high-resolution measurements on the $2\leftarrow0$, $3\leftarrow0$, $4\leftarrow0$, $5\leftarrow0$ infrared absorption bands of HI and DI, *J Mol Spectrosc* 85: 253–270, 1981.
53. Bulanin MO, Domanskaya AV, Kerl K, Maul C: Spectral line parameters in the $(3\leftarrow0)$ overtone band of the HI molecule and line-mixing in the band head, *J Mol Spectrosc* 230: 87–92, 2005.
54. Kiriyama F, Rao BS: Electric dipole moment function of $H^{79}Br$, *J Quant Spectrosc Radiat Transfer* 69: 567–572, 2001.
55. Tipping RH, Ogilvie JF: The influence of the potential-energy function on vibration–rotational wave functions and matrix elements of diatomic molecules, *J Mol Struct* 35: 1–55, 1976.
56. Ogilvie JF, Rodwell WR, Tipping RH: Dipole moment functions of the hydrogen halides, *J Chem Phys* 73: 5221–5229, 1980.
57. Morse PM: Diatomic molecules according to the wave mechanics. II. Vibrational levels, *Phys Rev* 34: 57–64, 1929.
58. Dunham JL: The energy levels of a rotating vibrator, *Phys Rev* 41: 721–731, 1932.
59. Makarewicz J: Energy levels of a perturbed Morse oscillator, *J Phys B: At Mol Opt Phys* 24: 383–398, 1991.
60. Sage ML: Morse oscillator transition probabilities for molecular bond modes, *Chem Phys* 35: 375–380, 1978.
61. Tyablikov SV: *Methods in the quantum theory of magnetism*, New York, 1967, Plenum Press.
62. Kolomiitsova TD, Burtsev AP, Peganov OP, Shchepkin DN: Absorption spectrum of the $(SF_6)_2$ dimer in liquid argon solution, *Opt Spectrosc* 84: 381–387, 1998.
63. Herzberg G: *Molecular spectra and molecular structure. II. Infrared and Raman spectra of polyatomic molecules*, Princeton, NJ, 1945, Van Nostrand.
64. Dong S-H: *Factorization method in quantum mechanics*, Dordrecht, 2007, Springer.
65. Green HS: *Matrix mechanics*, Groningen, 1965, Noordhoff.
66. Wei H: Four-parameter exactly solvable potential for diatomic molecules, *Phys Rev A* 42: 2524–2529, 1990.

www.ingramcontent.com/pod-product-compliance
Ingram Content Group UK Ltd.
Pitfield, Milton Keynes, MK11 3LW, UK
UKHW020125290926
15471UKWH00003B/295

9 780323 282666